Measuring Crime

Behind the Statistics

ASA-CRC Series on
STATISTICAL REASONING IN SCIENCE AND SOCIETY

SERIES EDITORS
Nicholas Fisher, University of Sydney, Australia
Nicholas Horton, Amherst College, MA, USA
Deborah Nolan, University of California, Berkeley, USA
Regina Nuzzo, Gallaudet University, Washington, DC, USA
David J Spiegelhalter, University of Cambridge, UK

PUBLISHED TITLES

Errors, Blunders, and Lies: How to Tell the Difference
David S. Salsburg

Visualizing Baseball
Jim Albert

Data Visualization: Charts, Maps and Interactive Graphics
Robert Grant

Improving Your NCAA® Bracket with Statistics
Tom Adams

Statistics and Health Care Fraud: How to Save Billions
Tahir Ekin

Measuring Crime: Behind the Statistics
Sharon Lohr

For more information about this series, please visit:
https://www.crcpress.com/go/asacrc

Measuring Crime
Behind the Statistics

Sharon Lohr
Professor Emerita of Statistics at Arizona State University

CRC Press
Taylor & Francis Group
Boca Raton London New York

CRC Press is an imprint of the
Taylor & Francis Group, an **informa** business

CRC Press
Taylor & Francis Group
6000 Broken Sound Parkway NW, Suite 300
Boca Raton, FL 33487-2742

© 2019 by Taylor & Francis Group, LLC
CRC Press is an imprint of Taylor & Francis Group, an Informa business

No claim to original U.S. Government works

Printed on acid-free paper
Version Date: 20190205
Printed by CPI Group (UK) Ltd, Croydon CR0 4YY

International Standard Book Number-13: 978-0-367-19231-0 (Hardback)
International Standard Book Number-13: 978-1-138-48907-3 (Paperback)

Library of Congress Cataloging-in-Publication Data

Names: Lohr, Sharon L., 1960- author.
Title: Measuring crime : behind the statistics / Sharon L. Lohr.
Description: Boca Raton, FL : CRC Press, 2019. | Includes index.
Identifiers: LCCN 2018055918| ISBN 9781138489073 (pbk. : alk. paper) | ISBN 9780367192310 (hardback : alk. paper) | ISBN 9780429201189 (ebook)
Subjects: LCSH: Criminal statistics.
Classification: LCC HV7415 .L64 2019 | DDC 364.072/7--dc23
LC record available at https://lccn.loc.gov/2018055918

Visit the Taylor & Francis Web site at
http://www.taylorandfrancis.com

and the CRC Press Web site at
http://www.crcpress.com

Contents

Preface

We all want less crime. But how can you tell whether crime has gone up or down?

Measuring Crime: Behind the Statistics helps you think statistically about crime. What makes some measures of crime more trustworthy than others? What features should you look for when deciding whether a statistic provides useful information?

The same ways of thinking can help you interpret statistics about unemployment, public opinion, time spent commuting, out-of-pocket medical expenses, high school graduation rates, seat belt use, smoking prevalence, automobile safety, and poverty rates.

You do not need a prior background in statistics or mathematics (beyond arithmetic) to read this book. There are no formulas or Greek symbols. The main ideas of statistical reasoning are discussed through examining the sources and properties of crime statistics. If you have taken one or more statistics classes, this book will give you a different perspective on the concepts you learned there.

You also do not need to be a professional statistician to investigate the data yourself. I describe examples of crime statistics and investigations that have been conducted on their quality, but do not present a comprehensive picture of crime in the United States. My goal is to help you evaluate statistics as a statistician would, so that you can find and interpret crime statistics on topics you are interested in.

The website http://www.sharonlohr.com has two online supplements with additional information. *Exploring the Data* tells how to obtain the data sources discussed in this book. To learn more about the statistical methods and examples, check *Endnotes and References*. It lists all of the sources used in the book, gives additional details and analysis for selected topics, and provides suggestions for further exploration.

I suggest reading Chapters 1 through 7 in sequence, as each depends on material in earlier chapters. Chapters 8 to 11 can be read in any order after that. Chapter 1 outlines the topics in each chapter.

I am grateful to the persons who read parts of the book and gave many helpful suggestions for improvement: Barry Nussbaum, Howard Snyder, Mike Brick, David Cantor, Rose Weitz, Norma Hubele, Lynn Hayner, and the series editors and reviewers for CRC Press. Jill Brenner and the Writers Connection group at the Tempe Public Library listened patiently on Friday afternoons to readings from a statistics book and reminded me to "show, not tell."

I'd also like to thank editor extraordinaire John Kimmel, who encouraged me to write the book and gently prodded me about deadlines, and the entire production team at CRC Press.

And finally, a big thank you to Doug, my partner in crime and everything else.

CHAPTER 1

Thinking Statistically about Crime

T HE HEADLINE of the February 2018 news story drew me in: "94%: Sexual misconduct in Hollywood is staggering." The story continued:

> The first number you see is 94% — and your eyes pop with incredulity. But it's true: Almost *every one* of hundreds of women questioned in an exclusive survey by USA TODAY said they have experienced some form of sexual harassment or assault during their careers in Hollywood.

Almost all of the women who participated in the survey said they experienced sexual harassment or assault. But what about women who did not participate?

We can't tell from this survey. The *USA Today* story acknowledged this: "As a self-selected sample, it is not scientifically representative of the entire industry."

Why don't the results from this survey generalize to all women who work in Hollywood?

- E-mail invitations for the survey were sent only to members of two advocacy organizations: The Creative Coalition, and Women in Film and Television. All survey participants have joined an organization that raises awareness about issues such as public funding for the arts and gender parity in the screen industries. Their experiences and perceptions likely differ from those of non-joiners.

- Altogether, 843 women participated in the survey. The story does not say how many were asked to participate, so we do not know what percentage of women responded to the survey invitation. Often, persons who choose to respond to surveys are particularly engaged in the topic or have strong opinions. The women who responded to the survey invitation may have had different experiences than the women who did not elect to participate.

A survey administered to a conveniently chosen sample is often cheaper and easier to conduct than a survey that is statistically representative. The *USA Today* survey provided timely context about fall 2017 news stories concerning sexual assault in Hollywood by asking a large number of people about their experiences. It demonstrated that the women who had come forward publicly were not the only ones with accounts of harassment and assault, and gave a voice to the survey participants.

However, the statistics from the survey apply only to the women who participated in it. It is correct to say that 94% of the 843 women who responded to this survey reported having experienced sexual harassment or assault—according to the survey's definitions and questions about those events—during their careers.

The study does not tell us what percentage of women in Hollywood have been sexually harassed or assaulted. That number might be 94%. It might be something else. The statistical procedures used in the survey do not allow us to assess how accurately the statistics describe *all* women in Hollywood.

STATISTICAL REASONING

Where do crime statistics come from, and how can you tell whether they are accurate?

Statisticians employ standard principles and procedures to collect data and to calculate and interpret statistics. This book illustrates how those principles apply to data sources and statistics about US crime rates, and tells you what to look for when evaluating a statistic.

The same principles apply to other statistics you encounter, from any field of study: political polls, unemployment rates, transportation use, size of bald eagle populations, agricultural production, diabetes prevalence, literacy rates, health care expenditures—the list goes on.

All statistics about crime, even the ones that appear to be exact counts such as number of homicides, are estimates. Statistical reasoning methods allow us to quantify uncertainty about estimates and tell how accurate they are likely to be.

WHY DO WE NEED ACCURATE CRIME STATISTICS?

You hear about crime all the time. Almost every edition of a newspaper contains at least one crime report. Stories about crime get high ratings—"If it bleeds, it leads"—and it is natural when you see one account after another to think that crime is everywhere.

Stories are memorable. But statistics tell us whether the stories are isolated events or reflect trends in society. When there are no high-quality statistics, people tend to extrapolate from personal experiences ("I know three people who were robbed last year—crime is really going up") and opinions.

Accurate crime statistics help answer questions such as:

- How much crime has occurred, and what types of crime are increasing or decreasing?

- Who are the victims and offenders?

- What are the costs of crime to victims and to society?

- What crime-prevention and crime-reduction strategies are effective?

- Where should law enforcement resources be allocated?

WHAT IS ACCURACY?

Any discussion of the accuracy of a statistic has to begin with the question: accurate compared to what? Suppose that there existed an omniscient statistician, who knows about every crime that is committed (even the so-called "perfect crimes" that are undetected), and knows the hearts and intentions of every perpetrator and victim. From a statistical point of view, a crime rate calculated by this omniscient statistician is about as good as one can possibly have. We'll call this rate the "true value."

But the true value depends on what is defined to be a crime. The *USA Today* survey described at the beginning of the chapter included nine types of experiences as harassment or assault, ranging from "having someone make unwelcome sexual comments, jokes or gestures about you" to "being forced to do a sexual act." Different definitions would have led to different answers.

A crime rate statistic has a numerator and a denominator—for example, a violent crime rate might be reported as 382 violent crimes per 100,000 inhabitants of the area. The definition used for crime determines the numerator of a crime rate statistic. Crime rates also depend on who is included in the denominator. Are children included? Prison inmates? Nursing home residents? Members of the armed forces? Some sexual assault statistics consider both sexes; others consider only women; others consider only women who are attending a college or university.

The set of persons or entities to whom the statistics are intended to apply is called the **population**. We would expect statistics about different populations to differ.

STATISTICAL PROPERTIES

Even if the same crime definitions and populations are used, crime statistics from different sources or samples are expected to vary. Almost every statistic deviates from the true value it estimates, usually for one or more of the following reasons:

Missing data. Almost all data collections fail to obtain some data, and for crime statistics this problem is of particular concern because often the missing data belong to crime victims. Undetected murders (such as deaths mistakenly ascribed to natural causes) will cause homicide statistics to be too low. Robberies not reported to the police will be missing from the law enforcement statistics for that crime.

Some surveys ask people about criminal victimizations they have experienced. If persons who are willing to answer the survey questions are more likely to be crime victims than those who decline to participate, then the victimization rate estimated from the survey data may be too high. The survey estimates may be too low if persons who agree to participate in the survey are less likely to be crime victims than persons who are asked to be in the survey but do not take part.

Measurement error. Measurement error occurs when an entry in the data set differs from the true value. Misclassifying a robbery as a purse-snatching is an example of a measurement error in police records. In surveys, measurement errors can occur because a question is worded confusingly or is misinterpreted, or because one interviewer might elicit a different response than another interviewer, or because the person responding to the survey does not tell the truth or has faulty memory.

Statisticians use the term "error" for anything that causes a statistic to deviate from its true value, but measurement errors should not be interpreted to mean "mistakes." Rather, they should be viewed as sources of uncertainty about statistics. Some measurement errors are indeed the result of mistakes, as when someone types the wrong value in the database, but others can occur simply because people interpret a question in various ways.

Sampling variability. Some crime statistics come from randomly selected samples of households, persons, businesses, or records. The statistic calculated depends on the particular sample that was drawn. If a different sample had been drawn, a different value of the statistic would have been obtained, and that leads to variability from sample to sample.

If you have taken a statistics class, sampling variability is probably the type of error you learned about—and it is often the only measure of uncertainty that is reported for crime statistics. But effects of missing data and measurement errors should also be considered when interpreting a statistic.

WHAT THIS BOOK IS ABOUT

This book is about the statistical ideas needed to interpret statistics about crime rates: definitions of crime, populations, missing data, measurement error, and variability.

These factors affect all statistics. The two national sources of homicide statistics—one set obtained from death certificates and the other from law enforcement agency reports—show parallel trends over time but have different numbers of homicides. Chapter 2 outlines some of the reasons for these differences, including different definitions, missing data, and classification error.

Law enforcement agency statistics on crimes such as assault and burglary are also estimates. The Federal Bureau of Investigation (FBI) collects and tabulates statistics from US law enforcement agencies on violent and property crimes. The statistic on the back cover about violent crime decreasing by 0.9 percent from 2016 to 2017 comes from the FBI's Uniform Crime Reporting System discussed in Chapter 3. The chapter describes crime classification errors and some of the statistical methods that can be used to measure and reduce them.

Of course, the FBI statistics include only crimes that are known to and recorded by the police. These statistics cannot tell us about crimes that are not reported to the police.

Surveys, however, can provide information on crimes not known to the police. They ask people about crimes that happened to them, and then ask whether they reported those crimes to the police. The US National Crime Victimization Survey (NCVS), the subject of Chapters 4 through 6, has surveyed US residents age 12 and older every year since 1973.

The NCVS is just one of many surveys that have been taken about crime. Some surveys give more accurate estimates than others. Chapter 5 explains why results from randomly selected samples can be generalized to a population and describes the procedure used to select the NCVS sample.

Chapter 6 describes the weighting methods used to try to compensate for missing data from persons who do not respond to a survey. It also discusses measurement errors in surveys: how do you ask questions and conduct the survey to elicit accurate responses?

Chapter 7 summarizes the statistical principles from the first six chapters, distilling them into eight questions that you can ask to assess the quality of any statistic you encounter. The remaining chapters of the book apply these ideas to two types of crime that are particularly challenging to measure (sexual assault and fraud), and look at how new data sources and procedures might improve crime statistics.

Sexual assault statistics from surveys depend heavily on what is counted as an assault, what questions are asked, and who asks the questions. Chapter 8 describes some of the experiments that have been conducted to study how different survey methods affect statistics about sexual assault.

Fraud and identity theft, the subjects of Chapter 9, were not included in the original list of crimes to be reported in the FBI statistics. In part because of this historical omission, the statistics available about fraud and identity theft are much sketchier than those about violent crime, burglary, and theft. Fraud presents a particular challenge because many victims are unaware they have been defrauded; others, such as persons in nursing homes, are excluded from many data collections.

Chapter 10 talks about the feasibility and challenges of using "big data," the massive amounts of data now available from financial transactions, social media, and police operations, to learn more about crime. Chapter 11 concludes the book with a historical perspective of US crime statistics along with some ideas for continuing to improve them.

The glossary at the end of the book defines terms and acronyms, and the website `http://www.sharonlohr.com` contains endnotes and links to data sources.

Although most of the examples are from US crime statistics, the statistical concepts apply to any data collection. The same statistical issues of operational definitions, populations, missing data, measurement error, and sampling variability affect almost any statistic you encounter.

Okay, let's get started. We explore homicide statistics in the next chapter.

SUMMARY

Crime statistics depend on how crime is defined, who provides data (and who does not provide data), how data are collected, what questions are asked, and how estimates are calculated.

A statistic can be thought of as

$$\text{Statistic} = \text{"true value"} + \text{"deviation,"}$$

where the deviation includes anything that causes the statistic to differ from the true value, either in a positive or negative direction. Statistics about crime rates can differ because the true values differ, the deviations differ, or both.

The true values depend on what crimes are measured, how the crimes are defined, and what populations are studied. Missing data, measurement error, and sampling variability cause a statistic to deviate from its true value.

Every statistic should be accompanied by a measure of uncertainty that assesses how close the statistic is likely to be to the true value.

CHAPTER 2

Homicide

HOMICIDE STATISTICS are usually reported as though the exact number is known: for example, "A total of 17,250 people were murdered in 2016."

In reality, all crime statistics are estimates. Although homicide statistics are thought to be more accurate than police statistics about other crimes (after all, not all robberies come to the attention of the police, but it is thought that most homicides do), they are still subject to measurement error and missing data. And, of course, they depend heavily on what is considered to be a homicide.

The US has two major data sources on homicide. Statistics in the FBI's Uniform Crime Reports (UCR) are compiled from data provided by law enforcement agencies; about 17,000 agencies provided data in 2016. The Centers for Disease Control and Prevention (CDC) statistics on homicide come from death certificates.

Figure 2.1 displays the estimates of national homicide rates from the FBI and CDC from 1980 to 2016. They show similar trends over time, but there are some differences. We explore statistical reasons underlying those differences in this chapter.

WHAT COUNTS AS A HOMICIDE?

The FBI's UCR program considers three types of homicide. *Murder and nonnegligent manslaughter*, the "willful (nonnegligent) killing of one human being by another," is the only type counted in the violent crime statistics. *Negligent manslaughter*, the "killing of another person through gross negligence," and *justifiable homicide*, which includes killings in self-defense and killings by a law enforcement officer in the line of duty, are tallied separately. Homicide

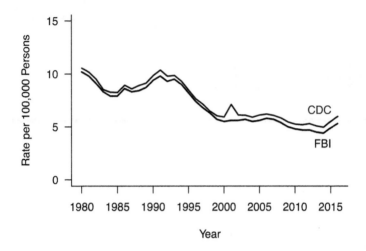

FIGURE 2.1 FBI murder rates and CDC homicide rates per 100,000 persons, 1980–2016.

is determined from the police investigation, not from judicial proceedings or investigation by a medical examiner.

The CDC statistics are tabulated from death certificates. Suspected homicides and suspicious deaths are referred to the office of a coroner or medical examiner, which then describes how the injury leading to the death occurred. The death certificates are forwarded to the Vital Statistics program, which assigns codes for manner and cause of death.

Some of the differences in the two lines in Figure 2.1 can be attributed to definitions. The FBI line displays statistics for murder, which includes only intentional killing. The CDC line includes intentional and negligent homicides, and thus would be expected to be higher than the FBI line.

Other differences come from the determination by different agents: law enforcement agencies for the FBI reports, and medical examiners or coroners for the CDC reports. The spike for 2001 in the CDC line, not found in the FBI line, occurs because the CDC counted the deaths from the September 11, 2001 terrorist attacks as homicides and the FBI did not.

WHAT IS THE POPULATION?

The statistics in Figure 2.1 display homicide trends as a rate per 100,000 inhabitants:

$$\text{homicide rate} = \frac{\text{number of homicides}}{\text{US population}} \times 100,000.$$

The population statistics for the denominator come from the US Census Bureau's annual reports of population, which are based on the decennial census statistics in census years and on intercensal population estimates in other years.

Statisticians usually compare time periods or geographic areas using homicide rates rather than numbers of homicides. While the CDC reported approximately the same number of homicides in 1984 (19,510) as in 2016 (19,362), the CDC homicide rate decreased from 8.3 homicides per 100,000 persons in 1984 to 6 homicides per 100,000 persons in 2016 because the US population increased during that time period.

The national population and state populations have standard definitions. But when comparing crime rates for cities over time, or from different data sources, make sure the statistics apply to the same population. Cities can annex adjoining areas over time. Law enforcement agencies can consolidate or disappear—for example, a town may decide to contract with another city for police services instead of staffing its own force—and new agencies can form.

In some areas, multiple law enforcement agencies have jurisdiction to investigate crimes and arrest offenders. FBI guidelines specify which agency should report an offense to the UCR, and which agency is considered to serve the area's population, so that offenses and people are not double-counted. But the guidelines result in some agencies being assigned a population much smaller than the population they actually serve, which can distort the agencies' crime rates.

The FBI and CDC use different rules for where to count a homicide. The FBI counts it in the state and city where the homicide occurred, while the CDC counts it in the state and county where the victim resided. Thus, a New York resident who is murdered in Vermont is counted as a Vermont murder in the UCR but as a New York homicide victim by the CDC. A tourist from Canada who is murdered in Vermont is counted in the UCR statistics for Vermont but does not appear in the CDC statistics.

The location-counting rule can cause the CDC and FBI homicide rates to differ for states and localities. A victim counted in the numerator of a state's CDC homicide rate is part of the state's resident population in the denominator. The population "at risk" for homicide is the denominator for the homicide rate.

In the UCR statistics, however, visitors to a city can be crime victims but are not part of the resident population used as the denominator for the crime rate. A city with a large number of tourists or commuters may have a higher UCR murder rate because the population "at risk" for murder is much larger than the number of residents in the denominator.

CLASSIFICATION ERROR

Misclassification is the primary source of measurement error for homicide statistics.

In the CDC data, a medical examiner may erroneously classify a homicide as a suicide, a natural or accidental death, or undetermined. Errors can also go in the other direction; for example, a medical examiner might record an accidental death in which a child is shot by another child as a homicide.

Misclassification occurs in police statistics when law enforcement officers make an incorrect determination while investigating a death. While errors can occur in either direction, it is likely that the net effect is to undercount homicides. A death suspected of being a homicide undergoes further investigation that can correct the initial impression, while a death initially reported as a suicide might not be investigated further.

The CDC does not use the term "justifiable homicide," but it has a separate classification, called legal intervention, for deaths caused by law enforcement actions. Killings by private citizens in self-defense are considered justifiable homicides by the FBI but are not considered legal interventions by the CDC.

Figure 2.1 excludes justifiable homicides and deaths from legal intervention; the different definitions and methods for determining these deaths explains part of the differences in the two lines. Again, misclassification may affect the statistics. Law enforcement agencies may classify some deaths as murders that are actually justifiable homicides, and vice versa. Sometimes the CDC statistics count a death caused by a law enforcement officer as a homicide instead of a legal intervention because the death certificate does not specifically mention that an officer was involved. We'll look at methods for estimating the number of deaths caused by law enforcement officers in Chapter 10.

Homicide statistics are reported to the FBI and CDC by states or localities within states. Different places may have different protocols for classifying uncertain deaths. More research is needed on how much variability is introduced into homicide (and other crime) statistics because the thousands of different medical examiners and law enforcement agencies may classify deaths differently.

Assessing Measurement Error

How can you tell how much measurement error affects a statistic? Typically, you need to perform experiments to see how statistics change with different methods of data collection, review records in a sample of the data, or compare with other data sources.

The FBI and CDC lines in Figure 2.1 track each other closely over time. This concordance from the two data sets provides confirmatory evidence that the homicide rate is lower in the 2010s than it was in the 1990s, even though the CDC rates are higher than the FBI rates each year.

Comparing the CDC and FBI statistics will not detect errors that affect both data sets similarly. Police departments and medical examiners' offices work together and usually adjudicate a death the same way. So-called "perfect" murders, where the death is not discovered or is not suspected to be a homicide, will be missed by both sets of statistics. (Homicides where the offender is unknown are counted in the statistics, even if the crime is never solved.)

INFORMATION ON HOMICIDE CHARACTERISTICS

When the UCR program began in 1930, law enforcement agencies were asked to submit monthly counts of the number of crimes of different types. Later, these submissions were coordinated through state UCR programs, but for most offenses the only data collected were agency summary statistics. There was no national data set containing details about individual incidents.

In 1962 the FBI began collecting data on individual homicide incidents through Supplementary Homicide Reports (SHR). The SHR data include information, when available, on the age, race, ethnicity, and sex of the victim and offender(s), relationship between victim and offender, circumstances of the homicide, and weapons used.

The CDC data, compiled from death certificates, contain information on the victim's age, race, ethnicity, sex, education, marital

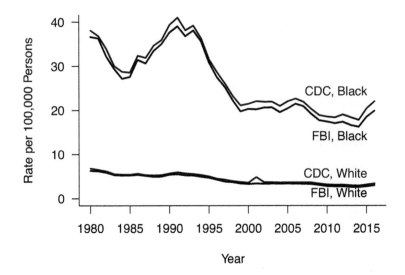

FIGURE 2.2 FBI murder rates and CDC homicide rates by race, 1980–2016.

status, residence, cause of death, and types of injuries. There is usually little or no information on the offender.

The types of additional information reflect the purposes of the data collections. The FBI data are collected to study crime and serve the strategic needs of the law enforcement community. The CDC data are collected to study and improve public health.

Measurement errors also occur for the additional information. The victim's race or age may be incorrect in either data set. The SHR may get the weapon or circumstances wrong (the SHR may say the murder occurred during a robbery when it actually occurred following an argument). Information on one or more characteristics may be missing, as when no age is recorded for the offender because the offender is unknown.

Errors and missing data on victim characteristics affect homicide statistics for age, race, and sex groups and other subsets of the data. Figure 2.2 shows that the difference between the CDC homicide rates and the SHR murder rates is larger for victims classified as black or African American than for victims classified as white, although the CDC and FBI statistics show similar patterns.

Evaluating Accuracy of Race Information on Death Certificates

The US Census Bureau's official population statistics on race rely on self-identification. You choose your own race and ethnicity categories on the census form and on government surveys.

Law enforcement personnel determine a homicide victim's race for the FBI statistics, and funeral directors determine it for the CDC statistics. Sometimes they ask relatives about the decedent's race, but sometimes they rely solely on their own observations. And sometimes they record a different race than the decedent would have chosen.

In Figure 2.2, the homicide rate for blacks is calculated as the estimated number of black homicide victims divided by the estimated number of black persons in the population. The population count in the denominator comes from self-reported race classifications; the homicide count in the numerator relies on the race determined by the law enforcement agency or funeral director. If they tend to misclassify some victims of one race as belonging to a different race, without compensating errors in the other direction, then the estimated homicide rate for that race will be too low.

Elizabeth Arias and colleagues investigated how well the race on death certificates matched self-reported race. They could not ask decedents their race, but they could discover what race some of them had reported while alive.

Arias and her team compared the race and ethnicity reported on a person's death certificate with the race and ethnicity that the person had previously reported on the Current Population Survey. The Current Population Survey has collected data about unemployment from large representative samples of households since 1940. Enough decedents would have participated in the survey to be able to compare the self-reported race and ethnicity with that on the death certificate.

While almost all blacks and whites had the same classification in the survey as on the death certificate, only 88% of the persons who reported they were Hispanic in the survey were listed as Hispanic on the death certificates. Only 51% of the persons who reported that they were American Indians or Alaska Natives in the survey, and 91% who reported they were Asians or Pacific Islanders in the survey, had the same race on the death certificate.

Arias's investigation considered deaths from all causes. If the same patterns hold for homicides, death certificate errors for race and ethnicity may cause the estimated number of homicides for

Hispanics, American Indians, Alaska Natives, Asians, and Pacific Islanders to be too small in the CDC statistics.

The same types of errors presumably occur in the FBI's SHR data but have not been as well studied.

MISSING DATA

The CDC statistics are based on all resident death certificates filed in the 50 states and the District of Columbia. The CDC data are thought to include information on almost all deaths that occur in the US. The annual reports claim that "more than 99% of deaths occurring in this country are believed to be registered."

This does not mean that the CDC statistics include every homicide. As we have discussed, some deaths that truly are homicides are misclassified under another cause of death. This misclassification causes the CDC statistics to miss some of the homicides.

The FBI statistics miss murders that an investigation mistakenly concluded were attributable to another cause of death. And of course they miss murders that do not come to the attention of the police.

Nonreporting law enforcement agencies. The FBI statistics have an additional source of missing data. Although most law enforcement agencies report crime statistics to the UCR program, not all do. The FBI reported for 2015: "Of the 18,439 city, county, university and college, state, tribal, and federal agencies eligible to participate in the UCR Program, 16,643 submitted data."

If you just summed the number of crimes from the agencies that reported data, the estimated annual crime counts would be too small because they would not include crimes from the nonreporting agencies. Moreover, totals in successive years would vary simply because a different set of agencies reported statistics, not necessarily because the number of crimes changed.

The missing data problem is actually more extensive than indicated by the agency participation. Some of the agencies counted as reporting to the FBI have missing data for some months of the year, so those agencies' annual crime counts are too small.

Estimating number of homicides using imputation. The FBI **imputes** values for the months and years in which agencies have missing data. Imputation replaces a missing value in a data set with an estimate. This section describes the simple imputation procedure that has been used in roughly the same form since 1958; although different imputation procedures will likely be implemented in the early 2020s, the general principles of imputation will still apply.

For an agency reporting between 3 and 11 months of data, the average number of murders per month is calculated from the months of data that the agency reported. That average is used to estimate the number of murders in each of the missing months. Suppose an agency reports 3 murders in January, 4 in February, and 5 in March, and does not report for the rest of the year. Then the average value of 4 murders is imputed for each month from April through December, giving an estimated total of 48 murders for the year.

Nonreporting agencies, and agencies that report two or fewer months of data, have all of their data imputed. All agencies within a state, reporting or not, are divided among groups defined by metropolitan status, population size, and type of agency. Crime rates calculated from the reporting agencies in a group are used to impute values for the nonreporting agencies in the group.

Suppose that the reporting agencies within a group serve an area with 500,000 people and have a total of 25 murders for 2015, giving an estimated rate of 5 murders per 100,000 persons. Then the annual number of murders for any agency in that group with two or fewer months of data is imputed to be 5 times the agency's population (expressed as a multiple of 100,000). A nonreporting agency serving 40,000 people is estimated to have 2 ($= 5 \times 40,000/100,000$) murders for 2015.

Assumptions for imputation.   Is it valid to put in your own values for missing data? Yes, but only if you have tried everything possible to obtain the real data values, and are completely transparent about which values you imputed and what procedure you used to impute. While using an imputed value is not as good as having the real data value, it is usually better than doing nothing.

If the FBI estimated the total number of murders by summing the counts from agencies that reported data, and ignoring those that did not report or reported for only part of the year, they would in fact be using a bad imputation procedure in which all missing data values are replaced by zero. The estimated total number of crimes would always be too low.

By replacing the missing values with estimates calculated from agencies with similar urbanicity and population size, the FBI hopes that the estimated total number of crimes is closer to the value that would be obtained if all agencies reported their data.

No matter what you do with missing data, you are making assumptions about how the missing values are related to the information you have. The FBI's imputation method assumes that: (1) for an agency with between 3 and 11 months of reported data, the murder rate for the missing months is the same as for the months reported, and (2) the murder rate for nonreporting agencies is the same as for reporting agencies in the same group.

In general, imputation is most accurate if the reporting and nonreporting agencies in the same group are as similar as possible. If urbanicity and population size are related to murder rates, then performing imputations separately for each group gives more accurate statistics than would be obtained if the national murder rate were substituted for the missing rate of each nonreporting agency.

Much is known about nonreporting law enforcement agencies and the populations they serve—for example, poverty rate, demographic characteristics, and mortality statistics. It is likely that some of these characteristics are related to crime rates, and that a procedure that exploited this additional information (for example, forming groups that are similar with respect to these characteristics) would produce more accurate imputations.

We do not know if the FBI assumptions are valid since we do not have the missing data, but statistical methods could be used to assess uncertainty in the estimates caused by imputing values for the missing data instead of measuring them. Various researchers have evaluated how different assumptions and imputation procedures would change the estimates.

NATIONAL AND LOCAL STATISTICS

The FBI and CDC publish national statistics about homicide. Both also produce homicide statistics for states; the FBI publishes statistics for cities, metropolitan areas, and law enforcement agencies when available.

FBI and CDC homicide rates for states and smaller geographic areas often exhibit larger differences than the national statistics. This is partly because the population and number of homicides are smaller: one additional homicide has a larger effect on the homicide rate for a population of 300,000 than a population of 300 million.

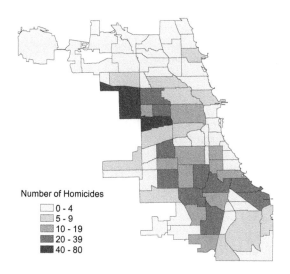

FIGURE 2.3   Homicides in Chicago community areas, 2017.

Different localities often have different types of measurement error. One agency may declare a homicide to be justifiable that another agency would declare as murder. An underfunded police department might not have the resources to investigate suspected homicides as thoroughly as it might wish. The idiosyncratic measurement error for one locality might not have a large effect on national statistics, but it can have a large effect on statistics for a small region.

Many states and police departments report and map crime statistics on their websites. These statistics are often updated daily and reported with more detail than the national statistics. Figure 2.3 displays a map of homicide counts for community areas in Chicago. Cities also publish maps and statistics for other types of crime and other geographic breakdowns.

The statistics from states and from police departments are subject to the same types of measurement errors as the national statistics. Crime counts from a police department's webpage should not be expected to correspond exactly to that city's UCR statistics. Crime definitions in state and local statutes may differ from the UCR definitions. In addition, maps on city websites display preliminary crime classifications, which can change.

When examining local statistics, beware of "crime increased by 50%" reports. Most community areas in Figure 2.3 had few homicides. The homicide rate doubles if an area with one homicide in 2017 has two homicides in 2018. Look at the population and the homicide count as well as the rate.

SUMMARY

The data for the two major sets of national homicide statistics—from the FBI's Uniform Crime Reports and the CDC's National Vital Statistics System—are collected for different purposes. The CDC data are collected to study public health, and the FBI data are collected to study crime. Because they use different definitions of homicide and data collection procedures, they produce different homicide counts and rates, although the national statistics show similar trends over time.

The FBI and CDC homicide statistics are often presented as though they were exact counts or rates, but both have errors from missing data and misclassification.

The CDC obtains information on almost every death that occurs in the US, but some deaths are placed in the wrong category: some homicides are ascribed to a different cause of death, and some deaths from other causes may be mistakenly listed as homicides.

The FBI statistics are likewise subject to misclassification, and miss homicides that do not come to the attention of the police.

Although measurement errors for homicide can go in either direction, the net result of the measurement errors and missing data is probably that the annual homicide statistics slightly undercount the true number of homicides.

If measurement errors and effects of missing data are consistent over time, however, the statistics accurately measure year-to-year changes in crime rates. If the estimated homicide rate for 2015 is, say, 18 units lower than the "true rate" for 2015, and the estimated rate for 2016 is 18 units lower than the true 2016 rate, then the estimated change, calculated as 2016 rate minus 2015 rate, will be the same as the difference of the true rates.

Some law enforcement agencies do not report data to the UCR program. The FBI uses imputation to estimate the number of murders for each nonreporting agency from relationships in the data provided by reporting agencies. Imputation introduces error because the estimated values can differ from the real data values.

CHAPTER 3

Police Statistics

H OMICIDE STATISTICS provide a measure of one type of crime, but what about other crimes? This chapter explores statistical properties of crime estimates from the FBI's Uniform Crime Reporting (UCR) program, which has collected counts of criminal offenses from states and law enforcement agencies since 1930. UCR statistics include only crimes known to the police.

The UCR is currently transitioning from the Summary Reporting System, which merely counts offenses, to the National Incident-Based Reporting System (NIBRS), which also collects data about each incident on the characteristics and relationships of victim(s) and offender(s), crime location (for example, a residence, grocery store, or park), weapon use, value of property lost, and other contextual information.

The target date for full transition to NIBRS is 2021. In 2015, 6,648 law enforcement agencies (about 1/3 of the agencies providing data to the UCR program) participated in NIBRS. The participating agencies represented about 30% of the US population and had, on average, different demographic characteristics from non-participants; consequently, NIBRS statistics from 2015 apply to the participating agencies but do not represent the entire country.

The statistical principles discussed in this chapter pertain to both systems of the UCR.

HOW ARE CRIMES DEFINED AND COUNTED?

The Summary Reporting System of the UCR collects monthly counts for the seven crimes in Table 3.1, plus arson and human trafficking. Rape, robbery, burglary, larceny-theft and motor vehi-

TABLE 3.1   UCR Summary Reporting System: Hierarchy of
Offenses

Offense	Definition
Criminal homicide	The willful (nonnegligent) killing of one human being by another
Rape	Penetration, no matter how slight, of the vagina or anus with any body part or object, or oral penetration by a sex organ of another person, without the consent of the victim
Robbery	The taking or attempting to take anything of value from the care, custody, or control of a person or persons by force or threat of force or violence and/or by putting the victim in fear
Aggravated assault	An unlawful attack by one person upon another for the purpose of inflicting severe or aggravated bodily injury
Burglary	The unlawful entry of a structure to commit a felony or a theft
Larceny-theft	The unlawful taking, carrying, leading, or riding away of property from the possession or constructive possession of another
Motor vehicle theft	Theft or attempted theft of a motor vehicle

NOTE: Definitions are quoted from the FBI website.

cle theft also include attempts to commit those crimes; an unsuccessful attempt to commit murder is an aggravated assault.

The list of crimes in Table 3.1 was formulated in the 1920s. The 1929 UCR manual included only crimes that were thought to be serious, amenable to collection via a standardized definition, and well reported to the police. In 2013, the FBI revised the definition of rape but otherwise the list is relatively unchanged.

Crimes not listed in Table 3.1—including fraud, identity theft, cybercrimes, stalking, vandalism, and simple assault (assault not meeting the definition of aggravated assault)—are not measured in the Summary Reporting System, although some of these are measured in NIBRS.

Hierarchy rule for counting offenses. In the Summary Reporting System, one offense—usually the highest in Table 3.1—is recorded per incident. An incident involving a rape and a motor vehicle theft is recorded as a rape and not as a motor vehicle theft.

Homicide, rape, robbery, and aggravated assault are violent crimes; an FBI statistic about the number or rate of violent crime includes these four, and only these four, offenses. The other crimes in Table 3.1 are property crimes.

Robbery is considered a violent crime because the offender is physically present with the victim and uses or threatens force. Burglary involves breaking and entering for the purpose of committing a felony or theft, without personal contact between victim and offender. If someone broke into your house while you were out and stole your television, you were burgled, not robbed.

Counting offenses with NIBRS data.   NIBRS asks law enforcement agencies to report each incident in as many offense categories as apply (up to 10).

The NIBRS data allow researchers to count offenses in multiple ways. The hierarchy rule can be applied to NIBRS data by counting only the most serious offense in each incident.

Alternatively, each of the separate offenses in an incident can be tallied. The FBI has estimated that between 8% and 15% of incidents involve multiple offenses, so counting each offense separately gives higher crime totals than using the hierarchy rule (except for homicide, at the top). An incident involving a rape and a motor vehicle theft contributes to the crime totals for each offense.

Which counting rule is better? It depends on what you want to know. If you want to tally only one offense per incident or to compare with historical trends, the hierarchy rule is appropriate. If you want to know how many motor vehicle thefts come to the attention of the police, it is better to count multiple offenses per incident—the hierarchy rule count excludes motor vehicle thefts that occur together with another offense.

But you should not compare a rate computed with the hierarchy rule from one year or jurisdiction to a multiple-offense-counted rate for another, because they measure different things. Similar warnings apply for comparing UCR statistics with those published by states and individual law enforcement agencies. Law enforcement agencies are asked to report statistics to the UCR program using the UCR classifications, but may use different classifications and counting rules for their own statistics.

As of 2018, some law enforcement agencies report their data through NIBRS and others report through the Summary Reporting

System; the FBI currently applies the hierarchy rule to the NIBRS data so that the crime statistics for NIBRS agencies are consistent with those from agencies that have not yet converted to NIBRS.

MEASUREMENT ERROR

Law enforcement agencies can only report statistics about crimes that come to their attention. Measurement error for a law enforcement agency's crime count is the difference between the agency's count and the count that would be obtained if every crime complaint known to the agency were correctly classified and recorded.

Crime classification errors.   Statistics for the crimes in Table 3.1 depend on how law enforcement agencies classify incidents. Measurement error occurs when a crime is placed in the wrong category or erroneously not recorded as a crime, or when a non-crime is wrongly included in a crime statistic.

Classification systems insist that each item be placed in one of a fixed number of categories but the distinctions among categories are not always clear. Just look at all the controversies there have been about classifying fossils into different species or classifying solar system objects as planets (poor Pluto was once a planet).

The UCR program publishes guidelines on how to classify crimes, but no publication can anticipate every situation and sometimes judgments must be made. Two police officers might in good faith interpret the same incident differently. Is a broken window the result of burglary (forced entry with intent to commit a felony or theft) or vandalism (malicious destruction of property, not counted among the crimes in Table 3.1)?

Persons making a complaint to the police may misremember or misrepresent an incident, or may lack the communication skills needed to provide the details required to classify the crime.

Some incidents are obviously aggravated assaults (for example, when a victim is shot or has severe injuries), but for others the decision between simple or aggravated assault is unclear. The classification makes a difference for the crime statistics of Table 3.1, where aggravated assaults are counted but simple assaults (assaults that do not involve use of a weapon and where the victim is not seriously injured) are not.

Variability across jurisdictions. Each of the more than 15,000 local police departments and sheriff's offices in the US works differently. There is bound to be variability in how they, and the more than 700,000 sworn officers, deal with crime complaints and report them to the UCR.

Data entry errors, coding errors, and data collection systems affect the estimates. In large police departments, reports often go through multiple levels of review and an error can be introduced at any of those levels.

Most of the errors discussed in this section affect the variability of statistics but should be approximately equally likely to occur in either direction. If you could repeat the year with the same incidents but with different law enforcement personnel, you would likely get a different burglary rate. But if the only measurement errors were random fluctuations and judgment calls that could go either way, you should get something close to the true rate if you averaged the results over all possible re-runnings of the year.

The next section discusses measurement errors that can lead to systematic and recurring differences between the published statistics and the values that should be reported.

SYSTEMATIC MEASUREMENT ERRORS

In the early 1990s, Grand Rapids, Michigan had the highest UCR rape rate among all cities with 100,000 or more people. Headlines called it "the rape capital of the United States."

That appellation was undeserved. It turned out that Grand Rapids had reported the figure for *all* types of sexual assault to the FBI, not just the crimes that fell within the narrow definition of rape used at the time. After the city revised its crime classification procedures, the rate dropped dramatically: from 215 rapes per 100,000 persons in 1992 to 59 per 100,000 persons in 1994.

Systematic measurement errors, unlike the errors discussed in the previous section, tend to recur in one direction (see Figure 3.1). They cause a statistic to be consistently larger (or consistently smaller) than the true value, making the statistic **biased**. Grand Rapids systematically overstated the number of rapes from 1990 to 1992 before correcting its reporting procedures.

Some systematic measurement errors come from procedures for reporting or confusion about crime definitions. Others may stem from incentives and numerical targets.

Systematic Error   Random Error

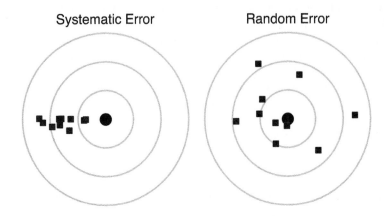

FIGURE 3.1 Systematic vs. random errors. The true value is the solid circle at the center.

Incentives and Numerical Targets

In the early part of the twentieth century, coroners were paid per death investigated. They received a fixed fee for each cause of death they established. But some localities required the fee, in case of murder, to be paid by the convicted murderer. If a stranger was found with a head wound, it was easier and more remunerative to declare the death accidental than to do a time-consuming investigation that might show the injury was caused by an assault. When the offender was unknown or unlikely to be caught, the coroner had strong financial incentive to declare the death due to something other than homicide.

Incentives, numerical goals, and quotas for work are used throughout society. Some incentive systems, though, can have unintended consequences. A customer service representative who is required to close 60 tickets each day may be motivated to find the fastest way to resolve each issue, even if the result is more unhappiness among the customers. A teacher whose continued employment is based on her students' test scores may emphasize test performance at the expense of long-term learning.

Incentive systems and arbitrary quotas can also affect the accuracy of data collected, and that is the concern here. People naturally want to present numbers that show themselves to advantage.

There is also evidence from many areas of life that some, when faced with a quota they cannot fulfill or an incentive system that rewards them for attaining a numerical target, will game the system or distort the numbers. Recent examples include allegations of teachers changing students' incorrect answers on tests, Veterans Administration hospital administrators altering the data on patient wait times, and bank employees opening fake customer accounts to meet sales quotas.

Incentives for police department statistics can be in either direction. Some cities set targets for crime reduction and hold personnel accountable for meeting those targets; in other places, having a high crime rate may provide an argument for more resources or grant funding.

How much does this sort of error affect UCR statistics? We do not know the extent, but multiple investigations have found evidence of systematic measurement error that may have been prompted by incentives. For example:

- A 2013 report found substantiated instances of downgrading (when a crime is recorded as a less serious offense) and suppression (when a person making a complaint to the police is discouraged from reporting a crime or no report is taken) of offenses in some precincts of the New York Police Department.

- A US Department of Justice audit found that a Georgia county's statistic of nearly 12,000 aggravated assaults, submitted in an application for a police officer hiring grant, was overstated by more than 10,000.

- The Office of National Statistics in the United Kingdom removed the "National Statistics" status—the certification that statistics meet quality standards—from police-recorded crime statistics in 2014 after allegations that those statistics had been affected by "a degree of fiddling."

Why does this matter? Flawed statistics not only present a misleading picture of crime, but can also lead to less effective policing. Suppressed crime complaints will not be investigated. Resources or grant funding may be directed toward areas with higher reported crime figures, even though other areas actually have more crime. An experimental program for police-community partnerships might be deemed ineffective for reducing crime when in fact it was highly successful. Community trust in a police department can decrease when flawed statistics are exposed.

METHODS FOR ASSESSING MEASUREMENT ERROR

As we discussed in Chapter 2, many statistical methods are available for studying measurement error in a law enforcement agency, although these are not guaranteed to catch all of it.

Compare statistics from your data with statistics from other data sources. Chapter 2 compared UCR and CDC homicide statistics, and Chapter 4 will discuss another, independent data source for crime statistics. You can also perform experiments to see how the statistics change with different data collection methods, as discussed in Chapters 6 and 8.

Perform statistical analyses of the data. A statistical analysis can identify agencies whose crime numbers or rates are unusual when compared with statistics from similar agencies, or when compared to predictions from a statistical model that relates crime rates to other characteristics. If most cities of a certain size have 10 times as many aggravated assaults as murders, a city whose aggravated assault rate is only twice as high as its murder rate might warrant further review.

This type of analysis does not prove there is systematic measurement error, because an unusual value may be accurate. It also may not detect systematic measurement errors that affect everyone, because in that case agencies with errors do not stand out. The FBI reviews numbers before they are published and checks on agency reports that seem unusual.

Have experts review a sample of records. A study of reports from 12 police departments in a southeastern state concluded that larcenies were overcounted while robberies, aggravated assaults, and burglaries were undercounted: UCR statistics from the departments had 1,179 violent crimes from the list in Table 3.1, while the experts' review of the narratives led to an estimate of 1,521 violent crime offenses.

Expert review of a sample of narratives is an example of an auditing approach. Major police departments often post results from internal auditing procedures. Audits of records can detect crime misclassification based on the written narratives, but not errors in the narratives themselves, or errors that occur because some complaints are not recorded in the database.

Ask personnel about types of errors. A national survey of nearly 8,000 police officers from large departments, conducted in 2016, estimated that 37% of officers reported formal or informal expectations to meet a predetermined number of tickets, arrests, citations, or summonses.

These methods can diagnose some of the sources of measurement error. But surveys and quality audits are expensive, and law enforcement budgets are limited. Too much review can lead to procedures in which the emphasis on producing and evaluating statistics detracts from an agency's mission to protect and serve.

Statisticians recommend assessing and reducing measurement errors through continual quality improvement programs. These programs modify data collection systems so that fewer errors occur in the first place.

STATISTICAL QUALITY IMPROVEMENT

The statistician W. Edwards Deming promoted a systems-based approach to achieving high quality. He argued that the biggest opportunities for improving quality lay in changing procedures and methods, not in focusing on particular individuals—improving the process as opposed to assigning blame. Replacing the persons who classify crime with new personnel will not improve statistics' accuracy if the classification instructions are unclear.

Before listing features of a quality improvement program, let's look at an example.

Los Angeles Crime Classification

The *Los Angeles Times* reported in 2014 that, based on the newspaper's review of the narrative descriptions in 94,000 police reports, the Los Angeles Police Department misclassified nearly 1,200 violent crimes—mostly aggravated assaults—as lesser offenses during a one-year period. The current and former police officers interviewed for the article gave differing reasons for the misclassification: some said it was inadvertent, while others said it stemmed from pressure to meet numerical targets for crime reduction.

Following the newspaper investigation, the Office of the Inspector General of the Los Angeles Police Commission initiated its own review. Its 2015 report estimated that 9% of cases classified as simple assaults between 2008 and 2014 should have been classified as aggravated assaults. Had all assaults been correctly classified, the aggregated assault rate would have been approximately 30% higher each year.

The review found that more than 70% of the aggravated assaults that had been mistakenly recorded as simple assaults fit in one of three categories:

1. Brandishing a weapon. The practice prior to 2015 was to count brandishments as simple assaults. But the UCR Manual stated that any incident in which a dangerous weapon "is used or threatened to be used" is aggravated assault. The review concluded that displaying a weapon counts as a threat to use it.

2. Strangulation and choking. Strangulations had usually been classified as aggravated assault only when the victim lost consciousness. After the review, the types of strangulation considered as aggravated assault were broadened to include those in which the victim had difficulty breathing or the offender threatened to seriously injure the victim.

3. Serious injury. Some of the misclassifications may have occurred because the California Penal Code defined serious injury more restrictively than the UCR. For example, in the California Penal Code a "wound requiring extensive suturing" is serious while in UCR guidelines an injury requiring any suturing at all is serious.

The report concluded: "While a small proportion of these mistakes appeared due to individual officer error, the majority appeared to be due to systemic failures of the system." It then recommended systems-level changes to improve data quality.

In response to the report, the police department directed that all incidents with crime code "Brandishing" should be coded as aggravated assaults, and issued revised classification guidelines with more detailed instructions for incidents involving strangulation and injuries. It also instituted training, a Help Desk, and an ongoing quality improvement program for crime data.

After these changes were implemented, the Office of the Inspector General found that fewer aggravated assaults were misclassified as simple assaults in the first quarter of 2015.

However, the number of aggravated assaults for Los Angeles in the UCR increased from 9,836 in 2014 to 13,713 in 2015. Part or all

TABLE 3.2   Possible System Changes for Improving Data Quality

1. Study the classification errors and identify areas where more training or different procedures could help.
2. Simplify the classification system so it is easier for people to classify crimes correctly.
3. Increase or improve training on collecting and recording data.
4. Reduce the number of levels of inspection for crime statistics. Deming stated that one can eliminate the need for mass inspection by building quality into the system.
5. Separate the data collection system from performance incentives.
6. Eliminate arbitrary numerical targets.
7. Implement a program of continual quality improvement. Quality improvement is a never-ending process, marked by repeated cycles of recognizing opportunities for improvement, testing proposed changes, reviewing the results, and taking action based on what was learned.

of this increase may have been the result of the revised crime classification procedures, which recorded more crimes as aggravated assaults.

Systems-Level Changes for Quality Improvement

The list in Table 3.2, by no means comprehensive, gives a few ideas for systems-level changes that some agencies have adopted (or could adopt). The Los Angeles review included items 1, 2, 3, and 7.

Numerical Targets in New Zealand

Why eliminate numerical targets for crime reduction (item 6 in Table 3.2)? Don't we want to have less crime? Yes, but Deming argued that, without a method to achieve the goals, setting arbitrary numerical targets accomplishes nothing and can actually harm quality. An investigation by Radio New Zealand illustrated Deming's point.

In 2012 the New Zealand government set ten numerical targets for the "Better Public Services" initiative. The crime-related targets included reducing the violent crime rate by 20%, the youth crime rate by 5%, and the reoffending rate by 25% by 2017.

By May 2017, however, violent crime had decreased by only 2%, not meeting the goal of 20%. That month the government announced a revised target of "10,000 fewer serious crimes by 2021," representing a 7% reduction from the projected 2017 figure.

The 2017 reoffending rate had not reached its target either. It had decreased by about 4%, far short of the target of 25%. The government "signalled it would change the way this was measured because the total number of reoffenders, as opposed to the rate, had dropped by 26 percent."

Deming said that setting arbitrary numerical targets harms quality and can result in false figures, especially if the system is not capable of meeting the goals. A better practice is to "work on a method for improvement of a process.... Only the method is important, not the goal. By what method?"

To be fair, the New Zealand program had specified some methods for reducing crime, including increased policing at high-crime locations and supporting repeat victims of crime. It had no protocol for evaluating whether those methods were effective, however. Even if violent crime had gone down, no one would have known what caused the reduction: it might have been due to the increased policing at hot-spots, or to other societal changes, or to decreased reporting of crime to the police, or even to random fluctuations.

But the numerical targets accomplished nothing. They led only to shifted goalposts when the original targets were not met.

Linking incentives to a narrow set of numerical targets can work against broader objectives. They may lead a department to focus on outputs that are easily quantified—for example, reducing the number of sexual assault complaints—at the expense of other desirable policing outcomes (perhaps the number of sexual assaults actually went down but the number of complaints increased because the community's trust in the police department has grown).

ANNUAL CRIME STATISTICS AND COMPARISONS

Annual crime reports based on police statistics almost always underestimate the total amount of crime that occurred. The UCR statistics report the number of crimes recorded by law enforcement agencies. They exclude crimes that never appear in those records.

Figure 3.2 illustrates how crimes in a law enforcement agency's jurisdiction make it (or not) into the agency's statistics, and hence

FIGURE 3.2 From crimes to police statistics. The vertical arrows show crimes that are not included in the police statistics.

into the UCR. The vertical blue arrows point to the crimes that are not included in the official statistics: they might not be recognized as crimes, or no one reports them to the police, or the police officers decide not to record them as crimes.

Crimes along the horizontal path in the diagram end up in the police statistics. Each successive box is smaller because of the crimes that are "lost" along the way. Even if a crime makes it into the rightmost box of Figure 3.2, however, it still might be classified as the wrong type of crime or be subject to other measurement errors.

If the measurement errors and effects of unreported crime are consistent from year to year or from place to place, then UCR statistics can track relative levels of crime.

For most types of crime, however, many factors other than the true amount of crime affect the statistics. Percentages of crimes reported to law enforcement agencies vary by crime type and across agencies. In addition, agencies and states have different procedures for dealing with crime complaints and recording statistics. These differences make it difficult to compare crime rates (with the exception of homicide) among cities using the UCR statistics.

Consider two cities, each with population 1 million. City A's UCR aggravated assault rate may be higher than City B's because it really has more aggravated assaults. But there are many other potential explanations for the difference. The actual number of aggravated assaults might be the same, but City A's residents are more likely to report them to the police. Or the two cities may classify assaults differently: City B calls an offense a simple assault that City A would classify as an aggravated assault.

Even within in the same city, estimated crime rates can increase (or decrease) for reasons other than a change in the true amount of crime. City A's rate may have increased from 2015 to 2016 because improved community relations led to a higher percentage of victims reporting the crime in 2016. The aggravated assault rate in Los Angeles increased after they changed the classification procedures to include more types of incidents in that category—this did not necessarily mean that there was more crime after the change.

Variability happens.   We expect crime rates to vary from year to year. Even if a city's homicide rate is fairly constant over time, the city will not have exactly the same number of homicides each year—it would be very surprising for it to have exactly 200 homicides every single year. In a small population, a small increase in number of crimes can lead to a large change in crime rates.

Missing data and NIBRS.   As discussed in Chapter 2, some law enforcement agencies do not report data to the UCR program, and that is likely to continue after the switchover to NIBRS in 2021. I expect that the FBI will impute missing data in NIBRS after it becomes the primary UCR data collection, and will provide measures of the uncertainty about the estimates that can be attributed to the imputation procedure.

No imputation was done for the NIBRS statistics for 2016 and earlier years. The NIBRS crime totals and rates for those years depend on the set of agencies reporting data, and that set changes from year to year. NIBRS crime totals have increased over time simply because a larger number of law enforcement agencies reported statistics in more recent years.

Even with partial data, however, NIBRS provides valuable information about victims and offenders that is not available from other data sources. For example, in every year from 1995 to 2016, about 80% of the violent crime offenders (among those where the offender's sex is known) have been male, even though different agencies reported data across those years. Some crime patterns are relatively unaffected by missing data, as we'll discuss in Chapter 6.

SUMMARY

The FBI Uniform Crime Reporting program in 2018 has two components: the Summary Reporting System, which reports crime counts alone, and the National Incident-Based Reporting System (NIBRS). The counts produced by the Summary Reporting System give little context for interpreting the crime statistics. NIBRS, when fully implemented in 2021, is expected to provide much more useful information about circumstances of crime as well as detailed information on victims, offenders, and their relationship.

UCR statistics estimate the numbers and rates of crimes recorded by US law enforcement agencies. They do not estimate the total amount of crime in the US, because many crimes are not reported to or recorded by the police.

The FBI emphasizes that UCR statistics do *not* measure the effectiveness of a police department. Many factors—for example, population size and density, economic and cultural conditions, climate, and residents' attitudes toward crime—affect the type and nature of crime in a community, as well as the number of crimes reported to the police. Although police departments play an important role in reducing some types of crime, they do not control all of these factors.

Measurement errors affect UCR statistics. Some measurement errors can be attributed to different interpretations of incidents or clerical mistakes, but others may be from systematic crime misclassifications.

If measurement errors and the percentages of crimes reported to the police are consistent over time, then UCR statistics track changes in the amount of crime. In other circumstances, however, a change in the crime rate can reflect a change in classification procedures or reporting, and does not necessarily mean that the underlying level of crime has changed.

Numerous statistical procedures exist for assessing measurement errors. These include comparing statistics from different data sources, performing statistical analyses to check for unusual data values, auditing records from the data, and asking personnel about types of errors.

Measurement errors can also be assessed by examining the procedures implemented to prevent them. The best way to ensure data quality is to build it into the data collection process. Statistical quality improvement methods can reduce measurement errors in police-reported crime statistics.

CHAPTER 4

National Crime Victimization Survey

L ONG BEFORE THE UNIFORM Crime Reporting (UCR) System began, it was well known that even if it worked perfectly, it would not capture all of the crimes that were committed. The UCR statistics include only crimes that the police observe or learn about from victims or witnesses.

There are many reasons why a crime victim might not report the incident to the police. A person may be unaware that he or she was a crime victim (for example, a child who was sexually assaulted); may be afraid or unwilling to report the crime (a domestic violence victim may fear reprisals or may not want to get the offender into trouble); may think the crime is too trivial to be reported or that reporting is futile; may not want to come to the attention of the police (a victim who was assaulted while involved in criminal activity); or may be embarrassed about or ashamed of being a crime victim (a person who was swindled out of life savings).

And the number of unreported crimes is quite large. In the US, it is estimated that only about 55% of serious violent crimes (rape or sexual assault, robbery, and aggravated assault) and 35% of property crimes (including burglary, motor vehicle theft, and theft but excluding fraud) came to the attention of the police in 2015. Motor vehicle theft typically has one of the highest rates of being reported to the police (usually greater than 75%), and is thought to be high because most insurance companies require a police report to reimburse a claim. The estimated percentages of these crimes

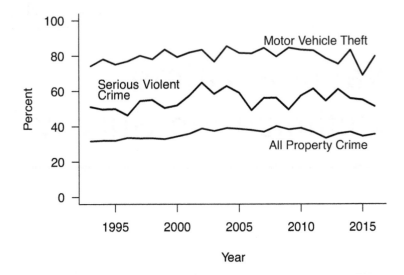

FIGURE 4.1 Estimated percentage of crimes reported to police, 1993–2016, from NCVS.

reported to the police have been fairly constant since 1993, with a few year-to-year fluctuations, as shown in Figure 4.1.

How do we know this? The statistics in Figure 4.1 are from the National Crime Victimization Survey (NCVS), which asks a representative sample of US residents about their recent experiences with crime. It provides detailed information on reporting to the police and consequences of victimizations that is unavailable from law enforcement reports.

Households and persons who provide information to the NCVS after being selected to be in the sample are said to have **responded** to the survey and are called **respondents**.

We'll look at how the NCVS works, statistically, in Chapters 5 and 6. In this chapter, we contrast the information we can get from a survey such as the NCVS with the information available from UCR statistics and police department databases.

DIFFERENCES BETWEEN UCR AND NCVS

Crime definitions. Because the NCVS depends on self-reports about crime, it has no information on homicide. NCVS definitions

TABLE 4.1   NCVS Crimes, from Most to Least Serious

Violent Crime	
Rape and sexual assault	Forced sexual intercourse; attack or attempted attack involving unwanted sexual contact.
Robbery	Completed or attempted theft, directly from a person, of property or cash by force or threat of force, with or without a weapon.
Aggravated assault	An attack or attempted attack with a weapon, regardless of whether or not an injury occurred, and attack without a weapon when serious injury results.
Simple assault	An attack without a weapon resulting either in minor injury or an undetermined injury requiring less than two days of hospitalization. Also includes attempted assault without a weapon and verbal threats of assault.
Nonviolent Personal Crimes	
Purse-snatching Pocket-picking	Theft or attempted theft of properties or cash directly from the victim by stealth without force or threat of force.
Property Crimes	
Burglary	Unlawful or forcible entry or attempted entry of a residence.
Motor vehicle theft	Stealing or unauthorized taking of a motor vehicle, including attempted thefts.
Theft	Completed or attempted theft of property or cash without personal contact.

NOTE: Definitions are quoted from the NCVS 2015 codebook.

for other crimes, listed in Table 4.1, differ from those in the UCR.

The published NCVS victimization rates count the most serious offense for each incident, using a hierarchy rule. However, the data sets contain information about other offenses in an incident so that multiple-offense counting rules may be used if desired.

Crime classification.   For the UCR, the law enforcement agency determines what type of crime occurred. If it deems the crime complaint to be valid, it records the incident in the summaries sent to the UCR program (see Figure 3.1).

NCVS survey respondents do not define the type of crime themselves. Instead, NCVS staff use the respondents' answers to questions to classify the type of crime.

The NCVS crime questions are asked in two parts. First, respondents are asked a series of "screening" questions that list examples of types of crimes that might have occurred, such as "Has anyone broken in or attempted to break into your home by forcing a door or window, pushing past someone, jimmying a lock, cutting a screen, or entering through an open door or window?" and "Has anyone attacked or threatened you ... with any weapon, for instance, a gun or knife? With anything like a baseball bat, frying pan, scissors, or stick?"

If a respondent answers no to all of the screening questions, the interview ends.

Respondents who answer yes to any of the screening questions are asked a second set of questions about the date and time each incident occurred, where it occurred (for example, home, school, work, parking lot, bus, hotel ...), whether the police were notified and what the police response was, what were the financial losses, what type of weapon was used, who else was present, who if anyone was injured, and other details.

The answers to the second set of questions are used to classify the type of crime. For example, an incident is classified as a completed burglary for the 2016 NCVS if the respondent gives the following pattern of answers to the second set of questions:

- "Were you or any other household members present when this incident occurred?" No.

- "Was there any evidence, such as a broken lock or broken window, that the offender(s) got in by force?" Yes, followed by a description of the evidence.

- "Did the offender actually get inside your [house/ apartment/ room/ garage/ shed/ enclosed porch]?" Yes.

The survey interviewer also records a narrative description of each incident, and this narrative is reviewed to make the final determination of type of crime.

Where incidents are counted.   The UCR counts a crime where it occurred; the NCVS (like the CDC homicide statistics) counts it at the victim's residence. The robbery of an Arizonan while visiting Florida is counted in Arizona for the NCVS and Florida for the UCR. The theft of a suburban commuter's car in downtown

Chicago is counted as a victimization of a suburban resident in the NCVS and as an urban crime in the UCR. Victimizations occurring in the US to non-US residents are not measured in the NCVS, but are included in the UCR if they are reported to police.

The location-counting rule makes little difference for national statistics, but it can cause UCR counts for cities with large numbers of commuters and visitors to be higher than NCVS counts of crimes reported to the police for those cities. The outsiders can commit or be victimized by crime, but are not counted in the resident population that serves as the denominator of the UCR crime rate.

Whose victimizations are counted. Anyone—any age or nationality—whose crime is reported to a law enforcement agency can be a victim in the UCR statistics. So can nonpersons such as businesses.

The NCVS statistics are limited to crimes against households and persons living in households and certain types of group dwellings (for example, rooming houses and college residence halls) in the US. Children under age 12 are not interviewed; nor are residents of institutions such as military barracks, prisons, hospitals, and nursing homes; nor are most homeless persons. Thus, the NCVS provides no information about crimes against children, persons in institutions, or commercial establishments.

What type of crime rate is reported. The UCR crime rates report number of offenses known to the police per 100,000 residents.

The NCVS victimization rates give the number of victimization incidents per 1,000 households (for property crimes) or per 1,000 persons age 12 and older (for violent crimes).

The NCVS annual victimization rate is larger than the percentage of households or persons who were victimized by crime during the year. The victimization rate counts multiple incidents for respondents who have been victimized more than once. For example, according to the 2015 NCVS, there were 18.6 violent crime victimizations for every 1,000 persons but only 9.8 out of every 1,000 persons had been victimized at least once by violent crime.

Missing data and measurement error. Some households and persons who are selected to participate in the NCVS cannot be reached or refuse to be interviewed, resulting in missing data for the survey.

Measurement error arises because the NCVS relies on respondents' answers to questions about their experiences with criminal victimization. Those are not always accurate. People misremember events or may distort the details, or some interviewers may be better at eliciting accurate information than others.

Figure 4.2 shows the steps an incident must go through to be reported among the NCVS statistics. First, the person must be selected to be in the NCVS and respond to the survey. This step automatically excludes crimes against children and persons who are incapable of participating in the survey.

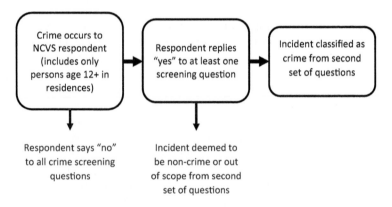

FIGURE 4.2   From incident to NCVS data.

Then, the respondent must interpret the incident as being a crime as described by the screening questions. These do not describe every type of crime that can possibly occur, and someone might respond "no" to all of the questions because he or she does not interpret an incident as belonging to one of the scenarios.

Finally, the respondent must give accurate answers to the follow-up questions for the incident to be classified correctly.

Just as some people do not want to report a crime to the police, some may not want to report an incident to an NCVS interviewer. For example, a domestic violence victim may not tell the interviewer about being assaulted because the abuser is listening during the interview, or a crime victim may simply not want to discuss the incident.

Measurement errors can also go in the opposite direction. Figure 4.2 shows a decreasing number of incidents through the interview steps, but the leftmost box starts with incidents presumed to be real. Some persons may exaggerate the severity of incidents when describing them to the interviewer, or even make things up (after all, no one verifies that the incidents happened as described). These types of measurement errors would cause victimization rates to be too high.

We'll discuss effects of missing data and measurement error on the NCVS in more detail in Chapter 6.

Scope of data collection.   The UCR attempts to obtain a count of crimes from every law enforcement agency. The NCVS collects information from a sample of persons and households. The next section discusses the implications of these different methods of data collection.

CENSUSES AND SURVEYS

The CDC homicide statistics discussed in Chapter 2 are based on a **census** of death certificates each year. The CDC Vital Statistics program collects information on every (well, almost every) death that occurs to a US resident. The UCR attempts to obtain information on every crime known to the police but does not quite succeed because some law enforcement agencies do not report their data. Still, it intends to be a census.

The NCVS is a **survey** of households and persons in the US. A sample of US households is selected to participate (more about how that is done in Chapter 5), and victimization rates are estimated from the information in the sample. In 2015, the NCVS conducted interviews with 95,760 households and 163,880 persons ages 12 and older within those households, out of the approximately 132 million households and 270 million persons in those age groups in the US population.

What are the advantages and disadvantages of taking a survey as opposed to a census?

Advantages of Taking a Survey Instead of a Census

A survey can be tailored to the topics of interest.   You can ask the questions you want to know about.

Many censuses are designed for administrative purposes, and might not have the type of information you can obtain when you determine your own questions. Although the CDC collects data on almost every death that occurs in the US, the information available about each death is limited. The program collects data on all deaths, not just homicides, and the questions on the standard death certificate reflect the multiple uses of the data. Different information would be requested if the only purpose of the data collection were to study homicide.

A survey can be less expensive. Interviewing people to ask them about their experiences with criminal victimization is expensive and time-consuming, and it is infeasible to do this for the entire population. Asking a sample of people, instead of the whole population, is much less expensive yet still produces reliable national estimates of victimization.

You can often have more control over the quality of the data in a survey. The survey-taker hires and trains the interviewers, establishes uniform procedures and standards for data collection, and takes other steps to assure the quality of the data. The same methods and procedures can be used for everyone who participates in the survey, giving more uniformity for the data collection.

By contrast, the CDC and FBI data collections on homicide and other crimes consolidate reports made by states and local agencies. The CDC and FBI provide guidelines and training for collecting data, and require states to have data quality programs, but they do not collect the data themselves.

You can measure the amount of error due to sampling variability in a survey. The **margin of error** of an estimate quantifies the uncertainty that is due to taking a randomly selected sample from the population instead of measuring everyone. You often see a poll result such as "53 percent of registered voters support candidate A" accompanied by a statement "the margin of error is plus or minus 4 percentage points." In general, statistics from larger samples have smaller margins of error.

Chapters 5 and 6 discuss margin of error and how to interpret it. For now, think of it as a standard measure that is used for one source of error—sampling variability—affecting an estimate. The margin of error does not include effects of missing data or measurement error.

TABLE 4.2 Number of Incident Records in 2015 NCVS

Crime Type	Number of Incident Records
All violent crime	1,386
Serious violent crime	531
Rape/sexual assault	103
Robbery	165
Aggravated assault	263
Simple assault	855
All property crime	5,809
Burglary	1,169
Motor vehicle theft	219
Theft	4,421

A survey can be more accurate than a census. Many people think that a census is more accurate than a survey because it tries to capture everyone. But, as we have seen in Chapters 2 and 3, censuses are still subject to missing data and measurement errors. The errors from those sources can far exceed the sampling variability in a carefully conducted survey sample. Moreover, we can quantify sampling variability, but often do not know the magnitude or direction of error caused by missing data and measurement error.

Disadvantages of Taking a Survey Instead of a Census

Survey sample sizes can be small for subpopulations. By the standards of most surveys, the NCVS collects an enormous sample, with more than 160,000 persons interviewed in 2015. By contrast, a typical political poll obtains data from between 500 and 1,500 respondents.

The large sample size results in low sampling variability and relatively small margins of error for NCVS estimates of national victimization rates. In 2015, for example, the rate of serious violent victimization was estimated as 6.8 victimizations (with margin of error plus or minus 1.2 victimizations) per 1,000 persons ages 12 and older.

But the majority of the respondents to the NCVS do not report any victimizations. This is a good thing from a societal point of view—it occurs because crime is relatively rare—but it means that the NCVS contains only a small number of records for some types of crime. Table 4.2 gives the number of records in the NCVS data file for each type of violent and property crime in 2015.

The 2015 NCVS data set contains 531 incident records of serious violent crime from the interviews with 163,880 persons. Any estimates that are calculated about the set of serious violent crime victims, such as the percent of victimizations reported to police used in Figure 4.1, are based on the responses in those 531 records.

Although the NCVS margin of error is relatively small for national estimates of victimization rates (calculated using all survey respondents), it is large when you want to estimate characteristics of crime victims (calculated using only the crime victims in the survey). It is also large for victimization rate estimates within subpopulations having small NCVS sample sizes, such as Hispanics age 18 to 24 or persons with disabilities.

The small sample sizes in Table 4.2 partly explain why the lines for percentages of serious violent crime and motor vehicle thefts reported to the police in Figure 4.1 are so wiggly. Each year's estimated percentage of property victimizations reported to the police has a margin of error of approximately 2 percentage points. But the estimated percentages of serious violent victimizations and motor vehicle thefts reported to police are based on smaller numbers of victimizations, and the margin of error for those statistics ranges from 6 to 10 percentage points each year. The year-to-year variability in Figure 4.1 can be explained by the small sample sizes.

By contrast, the 2015 National Incident-Based Reporting System (NIBRS) data from the UCR contained information on more than 178,000 motor vehicle thefts, 69,000 robberies, and 70,000 sex offenses. The NIBRS data are thus a tremendous resource for studying details of crimes.

But there is a trade-off: the NIBRS data are restricted to incidents known to law enforcement agencies that participate in the program, and have much less information about the details of the victimizations than the NCVS data. The NIBRS data have no information about victimizations not reported to the police.

Survey design and estimation can be complicated. It is easy to take a bad survey: just put up a web site and take the first people who respond. Surveys that provide reliable estimates are harder to conduct, and require a great deal of statistical expertise as well as scrupulous adherence to the survey procedures. Of course, censuses must also be conducted carefully, but often the protocols and estimation procedures are more complicated for surveys.

Missing Data in Surveys and Censuses

Remember the characterization of a statistic from Chapter 1 as a "true value" plus a "deviation." We can now expand this:

Statistic = "true value"
+ deviation due to sampling variability
+ effect of missing data
+ measurement errors.

A survey usually has two types of missing data. The first type is the data missing by design—any population member who is not selected to participate in the survey will not be in the survey data set. The errors from this type of missing data are included in the sampling variability. If we selected a different sample, we would get a different value for the statistic. The sampling error is equally likely to be positive or negative. Statisticians know how to measure the sampling error caused by taking a sample instead of a census through the margin of error.

The other type of missing data is from persons or entities who are selected to participate in the survey but cannot be reached or refuse to participate, called **nonrespondents**. This type of missing data also occurs in censuses and other data collections (often to a greater degree than in surveys). It is more worrisome than the first type because those who do not provide data can differ systematically from those who do. Moreover, we do not know how much the respondents and nonrespondents differ for the quantities of interest because we do not have the information from the nonrespondents.

UCR AND NCVS SERIOUS VIOLENT CRIME

We have seen that there are many potential sources of divergence for UCR and NCVS statistics: different definitions of crime, populations studied, methods of data collection, sources of measurement error, and missing data. How do these differences affect estimates of crime trends?

Let's look at the NCVS and UCR trends for serious violent crime: rape and sexual assault, robbery, and aggravated assault. These crimes are measured in both data collections. We would not expect the statistics to agree completely, because the UCR includes crimes against children and other groups out of the NCVS's scope, and the NCVS crime definitions differ from those in the UCR. But

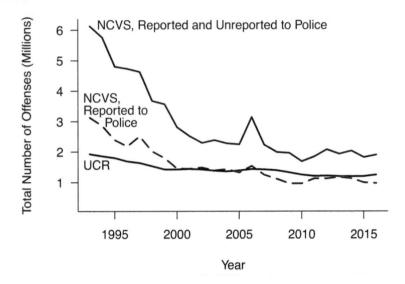

FIGURE 4.3 Estimated number of rapes, robberies, and aggravated assaults, 1993–2016, from NCVS and UCR.

we can still explore whether the two series exhibit similar ups and downs over time.

Figure 4.3 shows the estimated number of serious violent crimes from the UCR from 1993 to 2016. It also shows two estimates from the NCVS: the first gives the total estimated number of serious violent crimes (both reported and unreported to the police), and the second gives the estimated number of serious violent crimes that the respondents said were reported to the police. Because the UCR line includes only offenses known to the police, it might be expected to be closer to the second NCVS line than to the first.

For most years after 2000, the difference between the estimated number of offenses in UCR and the estimated number of victimizations reported to the police in NCVS is smaller than the NCVS margin of error. These independent data collections show similar trends over time.

The UCR and NCVS-reported-to-police lines do not agree as much as the CDC and FBI homicide trends from Figure 2.1. Part of the divergence is due to different crime definitions and measurement error in the UCR, as discussed earlier. But part of the divergence (including the NCVS spike in 2006) is attributable to the sample selection methods and error properties of the NCVS, as we shall see in the next two chapters.

SUMMARY

The NCVS provides a measure of crime that is independent of the UCR statistics. It asks persons in a sample of households about their experiences with victimization. The NCVS provides detailed information on the consequences of victimization such as financial losses, injury, lost days of work, or psychological distress.

A survey, such as the NCVS, collects data from a sample drawn from the population instead of the entire population. A survey can be designed to collect the types of information desired and can be less expensive and more accurate than a census (which attempts to obtain information on everyone in the population but may not succeed).

Because surveys do not sample everyone, they have sampling variability: the estimated number of victimizations would be different if a different sample of households were selected. In addition, survey data sets can have small numbers of records for some subpopulations. Surveys and censuses are both subject to measurement error and missing data.

CHAPTER 5

Sampling Principles and the NCVS

OW CAN STATISTICS FROM A SURVEY that you did not participate in reflect your experiences? The National Crime Victimization Survey (NCVS) estimates national victimization rates from interviews with fewer than 200,000 households. How does that work?

Consider the last time you had blood drawn for a routine medical test. A phlebotomist probably drew one to three small tubes of blood—she did not take every drop in your body. That little pump in your chest mixes the blood in your body so that a sample of it suffices for estimating your white blood cell count and cholesterol levels. The mixing makes the small sample **representative**—characteristics of your entire blood volume can be inferred from the sample.

The same principle applies to samples selected randomly from a population such as the set of US households. Random selection is like mixing up the population before taking a sample. **Probability samples**—that is, samples chosen using random selection methods—produce statistics that apply to the population from which the sample was drawn. Moreover, we can calculate how accurate the statistics are likely to be.

In some situations, non-randomly selected samples may have valuable information about population members. But it is difficult to assess the accuracy of the statistics from a non-probability or conveniently chosen sample. You have to assume that the survey participants are similar to those who did not participate.

In probability samples where everyone responds, you do not have to make any assumptions about the nonparticipants—the sampling procedure takes care of the assumptions for you.

SAMPLING WEIGHTS

Survey samplers use sampling weights to generalize from a sample to a population. The **sampling weight** of a unit in a sample tells how many units in the population it represents.

Let's see how weights work for a fictional example that allows us to explore the statistical principles without worrying about other complications. We'll return to the NCVS later in the chapter.

A city with 200,000 households wants to learn about residents' victimization experiences and attitudes on community safety. Police department statistics indicate that the South region, with 50,000 households, has higher crime rates (for crimes known to the police) than the North region, with 150,000 households.

The city plans to mail a questionnaire to a sample of 2,000 households, anticipating that about half will return it. We'll look at nonresponse for this example in Chapter 6.

To explore how sampling and weighting work when there is no nonresponse, let's look at a characteristic that can be ascertained for every address in the sample: whether the household lives in a multi-family structure (such as an apartment building) or a single-family dwelling unit. Here are two of the many designs that could be used to draw the sample.

Design 1: Simple Random Sample

In a simple random sample, every population subset of size 2,000 has an equal opportunity of being drawn as the sample. Consequently, each household in the city has the same probability, 1/100 (= 2,000/200,000), of being in the sample. We can think of each sampled household as representing 100 households in the city— itself and another 99 households that were not drawn for the sample; its sampling weight is 100.

The sampling weights are used to estimate what the 198,000 non-sampled households look like. The estimated population (see Figure 5.1) has 100 copies of the first household in the sample, 100 copies of the second household, and so on through the 2,000th household in the sample.

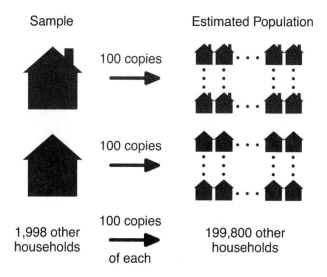

FIGURE 5.1   Design 1: Simple random sample.

All statistics are calculated from this estimated population. The total number of population households with a characteristic (for example, being in a single-family unit) is estimated by counting the number of households in the estimated population having that characteristic.

Of course, we do not create a physical data file with 200,000 entries. This is just a useful way to think about how weights work. Instead, we use the sampling weights directly for all calculations.

Rule for Weights: To estimate the total number of units in the population with a characteristic, sum the weights of the units in the sample with that characteristic.

Our simple random sample has 1,046 households that live in single-family units.

1. Estimate the total number of households in the population by summing all the weights: $2,000 \times 100 = 200,000$. (For simple random sampling, the estimate equals the population size.)

2. Estimate the total number of households living in single-family units by summing the weights of the single-family units in the sample: $1,046 \times 100 = 104,600$.

3. Estimate the percentage of households in single-family units by $100 \times (104{,}600/200{,}000) = 52.3\%$.

Percentages can be calculated directly from a simple random sample (1,046 out of the 2,000 sampled households live in single-family units), but in other sample designs the weights vary and in general we estimate a population percentage by

$$100 \times \frac{\text{sum of weights for sampled units with characteristic}}{\text{sum of weights for all sampled units}}.$$

Design 2: Stratified Sample

Any subset of 2,000 households can be drawn as the sample if simple random sampling is used. The number of South households in the sample depends on which sample was drawn—it could be 483 (the value in our sample), or 524, or 397, or even zero (although this last possibility is highly unlikely).

But we know the region for every household in the population, and can use this information to draw a sample that has a predetermined percentage of households from the South. Instead of taking a simple random simple of households from the entire city, we take one simple random sample from the North and a separate simple random sample from the South.

Selecting separate samples from different parts of the population is called **stratified sampling**. The two regions serve as the **strata** for this population. As long as we sample randomly from each stratum, we can reconstruct a valid estimate of the population from the combined sample and thus it is representative.

The sample size for each stratum can be set to any number desired. One can, for example, ensure that the sample mirrors the population's regional composition by drawing 1,500 households from the 150,000 in the North, and 500 from the 50,000 in the South, resulting in sampling weight 100 for each sampled household. Although the sampling weights are the same as for Design 1, this stratified design ensures that any sample drawn has the specified number of households from each region—a sample with 483 South households cannot be selected.

But we often want to take a stratified sample in which the sampling weights differ. Suppose the city wishes to compare estimates from the North and South regions while still being able to calculate citywide statistics. Then we might want to take the same sample size in the North as in the South, so that we have roughly the same amount of information for estimates from each region. That is what we do for Design 2.

Design 2, which specifies drawing 1,000 households from each region, **oversamples** the South. We draw 1 out of every 50 households (1,000/50,000) in the South but only 1 out of every 150 (1,000/150,000) in the North.

In the Design 2 sample, 600 of the 1,000 North households and 282 of the 1,000 South households live in single-family residences.

Sampling weights compensate for oversampling. But doesn't the oversampling make the sample unrepresentative? After all, we deliberately selected the sample so that it would have more households from the South region. If households in the South are less likely to live in single-family residences, then the raw percentage of the 2,000 sampled households that are in single-family residences (882/1,000, or 44%) underestimates the population percentage.

This is where the sampling weights come in. We don't calculate percentages directly from the sample, but use the weights to estimate a reconstruction of the population that has households from the two regions in their correct proportions.

The sample contains 1 out of every 150 households in the North region. Thus, each North household in the sample has sampling weight 150—it represents 150 households in the population. Similarly, each sampled household from the South represents 50 households in the population. A North household in the sample represents 3 times as many population households as a South household.

Figure 5.2 shows that even though households from the North region are underrepresented in the sample, the population estimated using the weights has the correct counts for each region.

Using the Rule for Weights,

1. Estimate the total number of households in the population by summing all the weights: $(1,000)(150) + (1,000)(50) = 200,000$.

2. Estimate the total number of households living in single-family units by summing the weights of the single-family-unit households in the sample: $(600)(150) + (282)(50) = 104,100$.

3. Estimate the percentage of households living in single-family units by $100 \times (104,100/200,000) = 52.05\%$.

Region	Sample		Estimated Population
North	1,000 households	150 copies ➡ of each	150,000 households
South	1,000 households	50 copies ➡ of each	50,000 households

FIGURE 5.2   Design 2: Stratified sample of 1,000 households from each region.

Oversampling some groups in the population is fine—*as long as you correct for the oversampling by using the sampling weights.* Statistics calculated without the weights cannot be said to apply to the population—they describe only the set of households that participated in the survey.

MARGIN OF ERROR

This discussion about weights still has not answered the question at the beginning of the chapter: How can we tell how accurate a statistic calculated from a sample is? That is, how close is our estimated percentage of households living in single-family units (52.3% for the simple random sample from Design 1, 52.05% for the sample from Design 2) to the percentage we would calculate from the 200,000 households in the population if we knew the single-family-unit status for all of them?

After all, we could apply weighting to any sample, regardless of how it is selected. We could, for example, use a list of 1,000 households belonging to a homeowner's association (all of whom live in single-family units) as the North sample, and assign each a weight of 150. That would give an inaccurate estimate of the population percentage of households in single-family units, but how would someone seeing the statistic know how far off it is?

Probability sampling allows us to say how accurate estimates are likely to be. We don't have to assume that the unsampled households are like the sampled ones. We can quantify how often we will draw a sample that will give an estimated percentage that is close to the population value, and how often we expect to draw a

TABLE 5.1   Margin of Error (MOE), in Percentage
Points, for an Estimated Percentage from a Simple
Random Sample

Sample Size	MOE if Estimated Percentage is		
	50%	20% or 80%	10% or 90%
100	9.9	7.9	6.0
500	4.4	3.5	2.6
1,000	3.1	2.5	1.9
2,000	2.2	1.8	1.3
5,000	1.4	1.1	0.8
10,000	1.0	0.8	0.6
100,000	0.3	0.2	0.2

"bad" sample—one in which the difference (statistic − population value) or (population value − statistic) exceeds a predetermined tolerance.

That tolerance is called the **margin of error** (MOE). Almost everyone uses the convention that 19 out of every 20 (95%) possible samples yields a statistic that is within the MOE of the population value.

Table 5.1 gives the MOE for estimating a percentage, say the percentage of households in single-family units, for simple random samples of different sizes selected from a large population. The MOE for an estimated percentage is largest when the percentage is 50%, and surveys often present the number in the 50% column as the MOE for all estimates since the MOEs for all other percentages will be smaller.

The MOEs in Table 5.1 apply only to simple random samples, but MOEs for other probability sampling designs such as stratified samples are easy to calculate using computer programs.

Our simple random sample of size 2,000 gave an estimate of 53.2% of households in single-family units, with MOE 2.2 percentage points. Had we randomly selected a different sample, we would have gotten a different estimate. There is a 95% chance that we have one of the samples where the range from (statistic − MOE) to (statistic + MOE) contains the true population percentage. For our sample, this range is from 50.1 to 54.5.

To think about this another way, suppose we look at 100 statistics from different probability samples (assuming no missing data or measurement error). We expect that about 95 of those statistics are accurate to within the MOE, and that the remaining 5 statistics are further away from the population value than the MOE. But we do not know *which* of the 100 statistics have the claimed accuracy.

There is a 1-out-of-20 chance that any particular simple random sample of size 2,000 is one of the samples in which the estimate is further than 2.2 percentage points—in either the positive or negative direction—from the population value. But even then we do not expect to be really far off. With a sample of size 2,000, about 1 out of every 100 possible samples will give an estimate further than 2.9 percentage points from the population value, and only 1 out of every million samples will give an estimate more than 5.5 percentage points away.

Four Points to Remember about Margin of Error

1. Taking a larger sample reduces the MOE.

2. The MOE does not depend on the size of the population (except when you sample more than 10% of the population). A simple random sample of size 2,000 has MOE of about 2.2 percentage points for a population of size 50,000, or 50 million, or 50 quintillion.

3. We do not know whether our sample is one of the "good" samples (where the statistic is accurate to within the MOE) or one of the "bad" samples where the statistic is further away. All we know is that 19 out of every 20 possible samples is one of the "good" ones.

4. The MOE measures sampling variability only. It does not include errors from measurement or missing data.

The last point is crucial for assessing the accuracy of surveys with nonresponse, and we'll talk about that in the next chapter.

Nonresponse, however, is often negligible in samples taken to assess measurement error.

LOS ANGELES CRIME CLASSIFICATION

The Los Angeles investigation of misclassified aggravated assaults (see page 29 of Chapter 3) employed a stratified sampling design. The population of interest was crime records that were not classified as aggravated assault but possibly should have been. These included simple assaults, weapons offenses, kidnapping, offenses against the family, and other crimes involving assault. How many of these records should have been classified as aggravated assault?

Each of the years under study, 2008 to 2014, had about 40,000 records for these types of crimes. Reviewing the narratives of all 280,000 reports to determine the correct classification would have been prohibitively expensive—and unnecessary for the goals of estimating the amount of misclassification and identifying types of incidents most likely to be misclassified. A sample of reports would provide enough information to improve the classification system and require far less effort.

The investigators used database information on incident characteristics to divide each year's records into two strata. Stratum 1 consisted of the crimes thought to have a higher chance of being misclassified because they had characteristics associated with aggravated assault (for example, aiming or brandishing a weapon or using a chokehold). Stratum 2 consisted of the remaining simple assaults.

Both strata needed to be sampled in order to estimate the misclassification rate. Although stratum 1 was thought to contain most of the misclassified incidents, investigators would not know whether that assumption is correct unless they also examined a sample of records from stratum 2.

About 2,500 cases per year fell into stratum 1, and the investigators selected a simple random sample of about 335 of those cases from each year to review. A simple random sample of about 220 records was selected from the approximately 37,000 records in stratum 2 for each year.

Is it fair to oversample the types of cases more likely to be misclassified when the goal is to estimate the misclassification rate? Yes—but only if the sampling weights are used to calculate estimates, as was done for all statistics in the report. Dividing the total number of misclassified cases in the sample by the total sample size (ignoring the weights) would give an estimated misclassification rate that is too high because misclassified cases had a higher chance of being selected for the sample.

The sampling weight for each sampled record from stratum 1 is approximately 7.5 (= 2500/335)—each case in the sample represents about 7.5 records in the population. The weight for each sampled record from stratum 2 is approximately 168 (= 37,000/220). The total number of misclassified records for each year was estimated by the sum of the sampling weights for the misclassified records for that year (approximately 7.5 times the number of misclassified records in stratum 1 plus 168 times the number of misclassified records in stratum 2).

The stratified sample gave an accurate estimate of the misclassification rate for each year (about 9% of cases were misclassified per year, with a margin of error of about 3 percentage points) but also yielded a larger number of misclassified incidents for study than a simple random sample would have.

HOW THE NCVS SAMPLE IS SELECTED

Although simple random sampling is the foundation of survey methods, few large surveys use it directly. These surveys cost millions of dollars, and no federal agency wants to end up with one of the rare but conceivably possible bad simple random samples that would look unrepresentative—for example, a sample where most of the households are in the western US, or a sample with almost no rural residents.

The NCVS uses stratified sampling to prevent these types of bad samples from ever being selected. It carries out the sampling in two stages, where the first stage selects counties to be in the sample and the second stage randomly selects housing units within the sampled counties. The NCVS then attempts to interview all persons age 12 and over in the sampled housing units.

A household's first interview with the NCVS is usually conducted in person at the household's address, and it costs a lot of money to reach a household and interview household members. Limiting interviews to a sample of counties reduces travel costs and makes it easier to manage the survey.

Sampling counties and housing units. The 2016 NCVS assigned each of the 3,144 US counties to one of 542 strata, then drew a probability sample of one or more counties from each of those strata. The stratification forces the sample to have counties from all regions of the country, from low-crime and high-crime areas, and from urban and rural areas. Households in counties that are not sampled in a stratum are represented by households from the sampled counties in that stratum.

Some counties are so important to the estimates that they are selected with probability one. If the sample for one year contained Los Angeles County but the sample for the next year did not, the estimated crime rates might differ merely because Los Angeles was in the first sample and not in the second. Thus, the NCVS sample always includes the counties in large metropolitan areas such as Los Angeles. This reduces the variability from sample to sample and hence results in smaller MOEs for survey estimates.

After the counties are selected, a probability sample of housing units is drawn from the counties in the sample.

Interviewing. After establishing contact with a sampled household, the interviewer asks a household member to provide information about the housing unit (for example, rented or owned) and all persons living in the household (age, race, ethnicity, sex, marital status, and education), and then asks about property crimes that occurred to the household in the 6 months preceding the interview.

Separate interviews are conducted with each household member age 12 and older to ask about victimizations he or she experienced personally in the last 6 months.

Residents of a sampled housing unit are interviewed 7 times, at 6-month intervals over a period of 3 years. If the family living at the address moves away during this period and new persons move in, the new family is asked to participate in the remaining interviews scheduled for that address. After 7 interviews, that address is removed from the sample and another randomly selected address from the same area takes its place. The first interview with a household is conducted in person, but subsequent interviews are usually conducted by telephone.

Respondents are told that all information is kept confidential.

Sampling weights in the NCVS. When the NCVS was launched in 1973, every housing unit in the US had about the same chance of being selected for the sample. This meant that each housing unit in the sample had approximately the same sampling weight, representing approximately the same number of housing units in the US population.

The 2016 NCVS design gives some housing units a higher chance of being selected than others. That change was made because of plans to produce separate victimization estimates, accurate to within a prespecified MOE, for each of the 22 most populous states. Remember that the MOE depends on the sample size (not the size of the population), so each of those states needs a sample that is large enough to meet the MOE requirement. Consequently, states such as Colorado (which has the smallest population among the 22) are oversampled—they are assigned a larger sample size than they would have had if every housing unit had the same chance of being in the sample.

The NCVS sampling weights account for this oversampling. Because housing units in Colorado have a higher chance of being in the sample than housing units from some other states, the sampled housing units from Colorado have smaller sampling weights. The sum of the sampling weights for housing units from Colorado in the NCVS sample approximately equals the total number of housing units in Colorado. The sum of the sampling weights for all housing units in the sample approximately equals the total number of housing units in the US.

The sampling weight for a housing unit selected to be in the NCVS sample tells how many US housing units are represented by that unit. But not everyone responds to the survey. The next chapter discusses how the sampling weights are modified to compensate for nonresponse.

SUMMARY

The NCVS and most other large government surveys employ probability sampling to select households and persons to participate in the survey. Random selection methods are used to obtain a sample that is representative of the population—that is, the sample can be used to estimate characteristics of the population and to calculate margins of error that show the accuracy of those estimates.

Each object or person drawn to be in a probability sample has a sampling weight, which tells how many population members it represents. Individuals with lower chances of being selected for the sample have higher weights to compensate. You can think of the sampling weight as telling how many times the sampled individual should be copied in order to reconstruct an estimated population from the sample.

The sum of the weights for objects in a sample with a characteristic estimates the number of population members with that characteristic.

The accuracy of an estimate calculated from a probability sample is given by the margin of error. This measures the uncertainty about the estimate due to taking a sample instead of measuring everyone in the population. The margin of error does not include uncertainty from measurement or missing data.

CHAPTER 6

NCVS Measurement and Missing Data

C RIME STATISTICS from the National Crime Victimization Survey (NCVS), like all statistics, are affected by measurement error and missing data. This chapter looks at those sources of error and the steps taken to assess and reduce them.

Before you start immersing yourself in all of the things that might cause NCVS statistics to differ from the "true victimization rates," however, remember that all data sources have errors. The point is not whether NCVS statistics are perfect—no statistics are—but whether they give useful information about crime. We'll put the errors in context at the end of the chapter.

The issues discussed in this chapter help explain why different surveys have different victimization estimates. The types of measurement errors and patterns of missing data in other surveys differ from those in the NCVS, as we shall see in Chapter 8 when we examine surveys that measure sexual assault. Learning about measurement errors and missing data in the NCVS will help you understand how these issues affect other surveys.

MEASUREMENT ERROR

The NCVS obtains information from a sample of households and thus has sampling variability: if a different sample were selected the estimates would be different. Even if every person in the country answered the NCVS questions about victimization experiences, however, its statistics would still not give the "true" amount of crime because of measurement error.

What types of measurement error affect the NCVS?

Question wording. Survey estimates depend on how well the questions capture the concept being studied. A seemingly simple question such as "Do you own a car?" can be interpreted in various ways. What if the household's vehicle is in your spouse's name? Does it count as ownership if you are leasing or making car payments? Is a pickup truck considered to be a car?

The NCVS defines burglary as "unlawful or forcible entry or attempted entry of a residence." But the NCVS does not measure burglary directly. It asks respondents a series of questions about crimes that might have occurred to them. The NCVS statistic for number of burglaries estimates the number of victimizations that would be counted if every household in the US population answered the survey questions about burglary—not necessarily the same thing as the "true" number of burglaries.

We know that question wording affects responses because the NCVS questions were revised in the early 1990s with the goal of eliciting more accurate answers from respondents. Experiments conducted to test the new questions showed how even slight changes in wording affect victimization estimates.

In preliminary testing, the revised questions had captured more victimization incidents than the old questions. If the NCVS had switched the entire sample to the revised questionnaire all at once, no one would be able to tell whether an increase in victimization rates after the switchover reflected a real increase in crime or occurred merely because the revised questions prompted more incident reports.

Instead, half of the 1992 sample was asked the old questions and the other half was asked the revised questions. From 1993 forward, the entire sample was asked the revised questions. In this way, the change in crime from 1991 to 1992 could be estimated using the half of the 1992 sample answering the old questions, and the change from 1992 to 1993 could be estimated using the half of the 1992 sample answering the revised questions.

The experiment allowed researchers to see how the revised questions affected crime estimates. Both questionnaires were administered during the same time period, and the areas receiving the revised questionnaire were chosen randomly. Thus, if the victimization rate estimates were higher for the persons answering the

TABLE 6.1   NCVS Screening Questions for Burglary

1973–1992 (Old)	1993–2016 (Revised)
Did anyone break into or somehow illegally get into your home, garage, or another building on your property?	Has somebody broken in or attempted to break into your home by forcing a door or window, pushing past someone, jimmying a lock, cutting a screen, or entering through an open door or window?
Did you find a door jimmied, a lock forced, or any other signs of an attempted break in?	Has anyone illegally gotten in or tried to get into a garage, shed or storage room?
Did anyone take something belonging to you or any member of this household, from a friend's or relative's home, a hotel or motel, or vacation home?	Or illegally gotten in or tried to get into a hotel or motel room or vacation home where you were staying?

revised questions than for those answering the old questions, that difference had to be caused by the question differences—the two groups were similar on everything else.

Table 6.1 lists the old and revised screening questions for burglary. The revised questions have additional cues intended to jog the respondent's memory: they specifically mention sheds, storage rooms, cutting screens, and pushing past someone.

The estimated burglary rate for 1992 using the old questions was 49 burglaries per 1,000 households. The estimated burglary rate for 1992 using the revised questions was 59 burglaries per 1,000 households—20% higher, just by asking slightly different questions. Violent crime estimates were higher with the more specific cues in the revised questions as well.

Question ordering and context.   Questions that come earlier in a survey can influence how respondents answer later questions. The earlier questions may prompt respondents to think differently about events or to remember an incident they had forgotten about.

An experiment performed on an early version of the NCVS randomly divided 12,000 sampled households in each of 13 cities into two groups. Group A was asked only questions about whether the respondent had been a victim of crime (the crime screening questions). The persons in Group B were also asked the crime screening questions, but first they were asked about attitudes toward crime such as whether they thought the neighborhood was safe and how they rated the performance of the local police.

Because the households were assigned randomly to Group A or Group B, one would expect the Group A and Group B respondents to have had similar victimization experiences. But in most of the cities Group B, primed by the attitude questions, had estimated rates of violent and property crime that were about 20% higher than Group A. It appeared that the attitude questions prompted respondents to remember more victimizations, particularly victimizations not reported to the police.

Memory of crimes.   The NCVS asks respondents about crimes that occurred in the six months prior to the interview. The statistics depend on how well the respondents remember such crimes. A respondent interviewed in October may have forgotten about the lawn tools stolen from the garage in June.

Another respondent interviewed in October may misreport a robbery as having occurred in May even though it actually occurred in March (more than six months ago). If this respondent was also interviewed the previous April, the interviewer can check whether this robbery was also reported in April. If it was, the robbery from the October interview will be dropped so that the incident is not counted twice. However, if the robbery is reported on the respondent's first interview, the interviewer cannot check notes from a previous interview for possible double-counting. The NCVS downweights incidents reported on the first interview to partially adjust for this overreporting.

Respondents may also forget details about victimizations they report to the survey—whether it was reported to the police, who else was present during the incident, or what was stolen.

Most NCVS respondents report two or fewer victimization incidents during their interview. But a few have experienced so many victimizations during the past six months that they cannot remember all of them distinctly; the NCVS records up to 10 victimizations for these respondents, but the actual number may be different than recorded. This type of measurement error does not affect estimates of the percentage of persons who were crime victims, but it can have a large effect on estimated numbers of sexual assault and intimate partner violence victimizations.

A respondent may be unaware that something was a crime, or conceal or falsify incidents. An assault victim may not view the incident as a victimization, or may be unwilling to discuss the incident with an interviewer. Other persons may want to get through the interview as quickly as possible and not mention crimes that would prompt more questions about details. Some may report events that are not crimes or incidents that did not really happen or did not happen as described.

Interviewers and interview circumstances may affect answers. Some interviewers may be more skilled at getting people to discuss victimizations than others. Or a respondent may feel more comfortable discussing events with one interviewer than with another. Although this causes variability in the estimates, interviewer skill and rapport are usually randomly dispersed across the sample.

Other errors associated with interviewers, however, tend to be systematically in one direction. The NCVS selected new samples of counties in 2006 and 2016. New interviewers were hired for the new counties, but previous investigations had found that less experienced interviewers tend to collect more incident reports, perhaps because they were more recently trained. Part of the victimization rate increases in 2006 and 2016 (see the spike for 2006 in Figure 4.3) might have been due to the influx of less experienced interviewers.

Performance incentives and numerical targets can affect crime statistics reported by law enforcement agencies, as discussed on page 26. They can also affect the number of incident reports obtained by interviewers. Before 2011, interviewers were evaluated primarily on the response rates they achieved, giving an incentive to obtain high rates. After refresher training programs and performance evaluation standards that focused on data quality were implemented, interviewers spent more time on the screening questions and the persons interviewed reported more crime incidents.

Circumstances of the interview also affect crime reports. Because the NCVS deals with sensitive topics, its protocols specify conducting interviews in private whenever possible. But between 2009 and 2013 more than half of the NCVS in-person interviews (more than three-quarters of those with persons age 12–17) were conducted in the presence of others. The rate of serious violent crime (rape or sexual assault, robbery, and aggravated assault) was significantly higher for the private interviews (12 per 1,000 persons) than the non-private interviews (9 per 1,000). Respondents may be more reluctant to report victimizations when someone else (perhaps the perpetrator) is listening.

MISSING DATA

In 2016, about 134,700 of the 173,300 eligible households in the sample responded to the NCVS, meaning that the household response rate was 78% (134,700/173,300). No responses were obtained from households at about 38,600 of the sampled addresses.

Although the NCVS attempts to interview all persons age 12 and over in the sampled households, some of those persons cannot be reached or refuse to participate. In 2016, interviews were conducted with about 84% of persons within responding households. The participation varied by age group. Nearly 94% of persons age 65 and older provided answers to the survey after their household had agreed to participate, but fewer than 75% of persons age 18–24 and fewer than 65% of persons age 12–17 responded.

Figure 6.1 shows the percentage of eligible households that responded to the NCVS for each year between 1993 and 2016. The "Person" line estimates the response rate for persons selected for the sample. For 2016, an estimated 78% of households and 66% of persons (84% of 78%) responded to the survey.

The NCVS had remarkably high response rates from its inception through 2010. Almost all surveys in the 2010s are experiencing higher nonresponse than in previous decades, however, and the NCVS is no exception.

And there can be still more nonresponse even when a person participates in the survey. An interviewed person can decline to answer one or more survey questions. In 2016, almost all of the NCVS respondents answered the crime screening questions. But about 3% did not provide a valid age, and 22% did not provide information on household income.

The decreasing response rates raise concern about possible **nonresponse bias**—that victimization estimates calculated from the survey may be too high (or too low) because respondents tend to have had more (or fewer) victimization experiences than the nonrespondents. The NCVS, like most major surveys, uses weighting to try to reduce nonresponse bias.

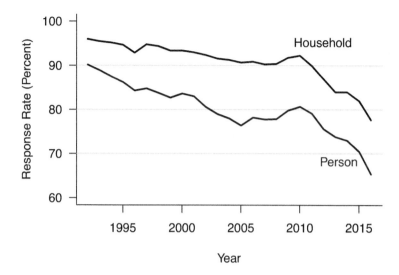

FIGURE 6.1 Household- and person-level response rates in the NCVS, 1993–2016.

ADJUSTING WEIGHTS FOR NONRESPONSE

Recall from Chapter 5 that the sampling weight tells the number of population units that a sampled unit represents.

Each nonresponding household has a sampling weight since it was selected to be in the sample. But a nonresponding household provides no data that can be carried over to the population households it is supposed to represent.

Most surveys redistribute the weight of a nonrespondent across a set of respondents that are thought to be similar to it. This section illustrates how that is done, and discusses when the procedure reduces nonresponse bias.

Weighting Adjustments: A Fictional Example

Design 2 from Chapter 5's fictional mail survey specified sampling 1,000 households from each region, North and South, in the city. To compensate for the higher chance that a South household is in the sample, the sampling weight for each North household is 150, and the sampling weight for each South household is 50.

But fewer than half of the sampled households responded to the survey. We were able to look up whether each household lived in a single-family residential unit, so we know that information for everyone in the sample. However, the characteristics we are really interested in, such as whether anyone in the household was a crime victim in the past 12 months, are known only for the respondents.

We can use the information known for everyone to divide the sample among four groups, called **weighting classes**: North households living in single-family units, North households in multi-family units, South households in single-family units, and South households in multi-family units. Table 6.2 shows the number of sampled and responding households in each class.

TABLE 6.2   Number of Sampled and Responding Households for the Four Weighting Classes

	Single-family unit	Multi-family unit
North	600 sampled households 400 respondents Response rate 67%	400 sampled households 180 respondents Response rate 45%
South	282 sampled households 143 respondents Response rate 51%	718 sampled households 216 respondents Response rate 30%

The lower response rates for households in the South, and for households in multi-family units, cause them to be underrepresented relative to households in the North and those in single-family units.

To correct for that underrepresentation, we distribute the sampling weights of the nonrespondents within each weighting class to the respondents in the class. Figure 6.2 illustrates this redistribution for respondents in North, single-family units.

The sampling weight of each respondent in a weighting class is multiplied by

$$1 + \frac{\text{sum of sampling weights for nonrespondents in class}}{\text{sum of sampling weights for respondents in class}}.$$

(The weight multiplier is expressed using the sum of the sampling weights rather than just the number of respondents because in most surveys the sampling weights vary within a weighting class.)

After the weight adjustment, each respondent represents its own share of the population plus part of the nonrespondents' share. The final weight, 225, estimates the number of population units represented by each responding household in the class.

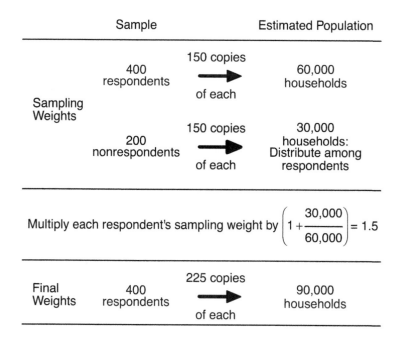

FIGURE 6.2 Weighting class adjustment for North, single-family unit respondents.

Table 6.3 shows the weight multipliers for all four classes. The calculations in the first column were done in Figure 6.2. The classes with the highest nonresponse have the largest multipliers.

Among the 400 responding North single-family-unit households, 36 reported that someone in the household had been a crime victim in the past 12 months; the sum of their sampling weights is $5{,}400$ ($= 36 \times 150$). After multiplying the sampling weight of each household in the class by 1.5, the sum of the final weights for the North, single-family-unit households reporting at least one victimization is $8{,}100$ ($= 1.5 \times 5{,}400$). This estimates the total number of households in that class that experienced at least one victimization in the past 12 months.

A similar calculation is done for each of the other weighting classes, as shown in Table 6.4. Using the final weights, the estimated percentage of households that experienced at least one victimization is $100 \times 22{,}346/200{,}000 = 11.2\%$.

Households from the South are underrepresented in the data because of that region's low response rate. But respondents in that

TABLE 6.3   Illustration of Weighting Class Adjustment Multipliers

Sum of sampling weights	North SF	North MF	South SF	South MF	Total
Sample	90,000	60,000	14,100	35,900	200,000
Nonrespondents	30,000	33,000	6,950	25,100	95,050
Respondents	60,000	27,000	7,150	10,800	104,950
Calculations					
Sampling weight	150.00	150.00	50.00	50.00	
Weight multiplier	1.50	2.22	1.97	3.32	
Final weight	225.00	333.33	98.60	166.20	

SF = single-family unit, MF = multi-family unit

TABLE 6.4   Respondent Households that Experienced at Least One Victimization

Households with 1+ victimizations	North SF	North MF	South SF	South MF	Total
Number of households	36	18	28	33	115
Sum of sampling weights	5,400	2,700	1,400	1,650	11,150
Sum of final weights	8,100	6,000	2,761	5,485	22,346

region were more likely to report that someone in the household had been a crime victim. When we use the final weights, we assume that the nonresponding households in that region were also more likely to have experienced crime.

When Do Weight Adjustments Reduce Nonresponse Bias?

Weight adjustments completely remove nonresponse bias for a survey question if nonrespondents within each weighting class would answer the question like the respondents in that class do. If nonresponding households are just as likely to have experienced a crime as the households in the same weighting class who respond, then a statistic calculated with the final weights is expected to be on target for estimating the population value.

If the respondents in a weighting class behave like a simple random sample of the people who were asked to participate in the survey, then they can be assumed to represent their share of the population in that class and the nonrespondents' share too.

In general, we don't know that the nonrespondents in a weighting class are like the respondents in the class; we need to assume that they are.

The weight adjustments reduce the nonresponse bias if the assumption is even approximately true. If, within each weighting class, nonrespondents' crime experiences are more similar to the respondents in that class than to respondent households in other weighting classes, then estimates calculated with the final weights are probably closer to the true values than estimates calculated with the sampling weights.

How can we tell if the weight adjustments improve the estimates? We can perform experiments and investigations to study possible effects of nonresponse—Chapter 8 describes an experiment on an e-mail survey about sexual assault—but the only sure way to know if the nonrespondents are like the respondents is to obtain their data.

We can also study effects of nonresponse by comparing statistics from the survey with statistics from an independent, high-quality data source. For example, we might compare average household size for our sample with that from US Census Bureau statistics—a large difference gives reason to be concerned that the crime estimates might be off as well.

Margin of error when there is nonresponse. Remember from Chapter 5 that when a large probability sample is selected and everyone responds, estimates calculated using the sampling weights will be accurate within the margin of error (MOE) in 95% of the possible samples that could be drawn. The result about the MOE is guaranteed when we use a probability sample; we do not have to assume anything about the households not sampled.

The MOE reported for a survey with nonresponse is calculated assuming that the nonrespondents are like the respondents in the weighting class. If that assumption is true, then the MOE of the estimate calculated with the final weights is a good measure of its accuracy. If nonrespondents are more likely to be crime victims than the respondents in the class, then the estimate calculated with the final weights underestimates the crime rate and the MOE, calculated assuming that nonrespondents are like the respondents, will understate the error in the estimate.

As a general principle, it is better to take a smaller sample, and invest resources to obtain a higher response rate, than to have a large sample but a low response rate. The high-response-rate survey has fewer nonrespondents about whom assumptions need to be made.

Weighting and imputation. Weighting and imputation (see Chapter 2) both attempt to compensate for nonresponse by using information from respondents who are thought to be similar to the nonrespondents. Some statisticians prefer weighting adjustments; others prefer imputation. Either method can reduce nonresponse bias if the respondents and nonrespondents in a weighting class (or a group used for imputation) are similar.

Weighting to External Counts

In some surveys little is known about the nonrespondents in the sample. Weighting adjustments can still be used if we know the number of people or households in each weighting class from an external source. For example, US Census Bureau statistics (considered to be the best available) give the total number of people in different age/race-ethnicity/sex groups and the number of households that are owner-occupied or renter-occupied.

Then the weight multiplier for each responding household in a weighting class is

$$\frac{\text{number of households in class, from Census Bureau statistics}}{\text{sum of sampling weights for responding households in class}}.$$

The sum of the final weights in a weighting class equals the Census Bureau's count for that class.

If nonrespondents in a weighting class are similar, on average, to the respondents in that class, then estimates calculated using the final weights are likely to be closer to the true population value then estimates calculated using the sampling weights.

Most survey samplers use numerous weighting classes so that the sampled households or persons in a weighting class are as alike as possible, at least with respect to the characteristics used to form the classes. For example, a weighting class might consist of black men age 65 and over with a college degree. The nonrespondents in that class have the same race, sex, age group, and education status as the respondents; the sampler hopes that they will also have had similar victimization experiences.

Weight Adjustments in the NCVS

There is a lot of evidence that members of some groups are, on average, more likely to respond to the NCVS than others. The response rate is higher for women than men, and higher for older persons than younger persons.

There is also a lot of evidence that victimization rates differ across groups. Every year since the NCVS began, the violent crime victimization rate has been more than five times as large for persons age 12 to 24 as for persons age 65 and over.

The age groups with the lowest response rates have the highest violent victimization rates. If the NCVS did nothing to try to compensate for nonresponse—if it estimated violent victimization rates using the respondents' unadjusted sampling weights—the estimates might be too low because the nonrespondents are disproportionately in high-victimization-rate age groups.

In the NCVS, each household's sampling weight is determined as described in Chapter 5. The 2016 NCVS uses multiple steps to calculate the final weight for each person responding to the survey. It is a complicated process, and this list, for simplicity, omits some of the steps (such as adjusting the weights for sampled counties, and adjusting for the possible overreporting of crimes by respondents on their first interview).

1. Start with the household's sampling weight—the number of population households that it represents.

2. Adjust for household nonresponse.

   Weighting classes are formed using combinations of state or region, rural/urban/suburban status of residence, race of the person who would answer the household questions (race is imputed for the nonrespondents), and other characteristics. The weights from nonresponding households in each weighting class are redistributed to the responding households in that class so that they represent their share of the population plus part of the share of the nonrespondents in the class.

3. Adjust for persons in responding households who do not complete a personal interview.

   The interviewer obtains a roster of household members in each responding household. Each person is initially assigned the household weight calculated in Step 2. But some of the persons do not respond, so weighting classes are formed using state or region, age, race/ethnicity, sex, and household composition, and the weights of the responding persons in those classes are increased.

4. Adjust for differences between sample characteristics and those known for the population.

   After the weight adjustments for households and persons in steps 2 and 3, there are still differences between NCVS estimates of the numbers of people in demographic groups and the population estimates produced by the US Census Bureau. The final steps in weighting adjust the weights of respondents so that for each large state and for each census region (Northeast, Midwest, South, West), the sum of the weights for respondents in each sex, race/ethnicity, and age group category equals the Census Bureau's population estimate for that category.

   Respondents in groups with low response rates (such as persons age 12–17) end up with higher weight multipliers than respondents in groups with high response rates (such as persons age 65 and over).

Because victimization information is unknown for the nonrespondents, we do not know how much nonresponse bias remains after the weight adjustments. The nonresponse adjustments likely reduce the bias, but it is possible that the respondents and nonrespondents within weighting classes have had systematically different victimization experiences. Much more research needs to be done on methods for increasing response rates and for assessing and reducing potential effects of nonresponse bias in the NCVS.

PUTTING ERRORS IN CONTEXT

After reading page after page about the possible errors in the NCVS and other surveys, you might be ready to wonder whether any statistics should be trusted. Please do not do this.

Remember, statistics are collected to learn something that could not be learned otherwise. All statistics have errors. But by evaluating how accurate the statistics are likely to be, we can determine their usefulness for answering questions about crime.

Let's look at how measurement methods and missing data affect uses of NCVS data.

NCVS Errors and Margin of Error

The NCVS MOE measures uncertainty about the estimates from sampling variability, that is, the error caused by taking a sample instead of a census of the population.

The MOE does *not* include uncertainty about the estimates from measurement errors or nonresponse. Thus, it does not include the uncertainty in the estimated victimization rates that results from failing to obtain responses from more than 20% of households. Survey-takers hope that the weight adjustments remove much of the nonresponse bias, but they do not know for sure.

A survey-taker can reduce the sampling variability and MOE by taking a larger sample size. But a larger sample size does not help with measurement errors or nonresponse bias. Even if the NCVS attempted to interview every household in the country, the measurement errors and nonresponse would persist—in fact, errors from these sources would likely worsen because there would be fewer resources available to ensure careful measurement or to follow up with households that are reluctant to participate.

The only ways to control measurement or missing data errors are to reduce them through continual quality improvement of survey procedures, or to attempt to compensate for them through weighting, imputation, or another statistical procedure.

Nonresponse May Affect Estimates of Relationships Less than Estimates of Counts or Rates

Nonresponse adjustments to the weights can have a big effect on estimates of victimization counts and rates. Subgroups of the population with lower response rates get larger weight inflations.

But nonresponse and the weight adjustments may have less effect on correlations and comparisons of victimization rates among subgroups than on estimates of victimization counts and percentages for the whole population.

Relationships among variables may reflect a "universal truth." Households with higher incomes tend to have more expensive houses than households with lower incomes. National Basketball Association players are taller than most other people. Smokers have shorter expected lifespans than nonsmokers.

Some of these relationships may hold among both respondents and nonrespondents to a survey. When that occurs, the relationship will be found when looking only at the respondents. In addition,

when the relationship is universal, estimates for relationships between variables calculated using weights may be similar to those
calculated without the weights.

There is another reason that relationship estimates may be less
affected by nonresponse. Suppose you want to compare the violent
victimization rate for adults who have a college degree with the
rate for adults without a college degree. If, after accounting for
the other variables used in the weighting, college graduates are
more likely to respond to the NCVS, then controlling for college
graduation status acts like an additional weighting adjustment.

Annual Victimization Rates vs. Changes Over Time

Errors from measurement methods and missing data may affect
estimates of year-to-year changes in victimization rates less than
estimates of victimization rates for a particular year.

Measurement error. We saw at the beginning of the chapter
that changing the NCVS screening questions on burglary resulted
in higher burglary rate estimates. Question wording clearly affected
the estimated victimization rates.

But, once the NCVS switched to the revised questionnaire in
1993, everyone was administered the same set of questions across
the country and every year. If the revised questionnaire fails to
capture some burglaries, it fails to capture them consistently.

As a consequence, comparisons of the burglary rate across regions of the country, across subpopulations, and across years are
likely to reflect true differences. The NCVS can measure whether
burglary increased between 2014 and 2015 because the same questions are asked both years.

Something similar happens for many of the other measurement
errors. Yes, people forget about some of the victimizations that
occurred, but on the whole they are usually no more likely to forget
about them in 2015 than in 2014.

You can tell whether crime went up or down as long as the
measurement errors are in the same direction and approximately
the same magnitude from year to year. Then, when looking at
changes in victimization rates, the measurement errors cancel.

However, if measurement error levels vary across areas or years,
then differences in estimated victimization rates might be at least
partially attributable to the measurement errors. In 2016, for ex-

ample, when the NCVS had a higher-than-normal percentage of new interviewers and made other changes in procedures, the survey report warned that the changes in estimated victimization rates from 2015 to 2016 might be due to the new procedures rather than a real change in the amount of victimization.

Nonresponse.   If the nonresponse bias that remains after weight adjustments has the same magnitude and direction each year, it too has less effect on estimates of change than on annual victimization rates. This was likely the case for the NCVS through 2010.

However, the sharp decrease in response rates in the 2010s raises concerns. In particular, the response rate among persons age 12 to 24—the age group that typically has the highest violent victimization rates—has dropped dramatically. This makes it more difficult to tell whether changes in victimization rates for this age group are from real changes in the true victimization rate, or because victims are becoming more (or less) likely to respond to the survey.

SUMMARY

The margin of error reported for the NCVS does not include uncertainty from measurement methods or missing data. Both of these affect estimates of victimization numbers and rates.

Question wording and context, respondents' memory and reporting of events, and circumstances of the interview all affect answers to the survey. Measurement errors can be assessed through experimental studies and comparisons with results from other data collections.

From 1973 until 2011, the NCVS household response rate was above 90%. Since then, however, the response rate has been dropping, raising concern about effects of missing data on the estimates.

The NCVS, like most large surveys, uses weight adjustments to compensate for missing data. It increases weights of respondents in weighting classes so that they represent their share of the population as well as the share of the population that was supposed to be represented by the nonrespondents.

Many states and cities conduct their own victimization surveys. These usually have much smaller sample sizes and lower response rates than the NCVS. The next chapter discusses how to evaluate other surveys according to the statistical principles discussed in Chapters 5 and 6.

CHAPTER 7

Judging the Quality of a Statistic

A N ARTICLE IN the February/March 2018 issue of *AARP: The Magazine* contained several statistics about fraud. It stated:

> Roughly 6 out of 10 cases of elder financial abuse are committed by relatives, according to a large-scale 2014 study. And about 3 out of 10 instances can be traced to friends, neighbors, or home care aides. In other words, 90 percent of perpetrators of fraud are known to their victims.

How do you evaluate a statistic you encounter? Table 7.1 lists eight questions you can ask, and this chapter works through those questions for the AARP statistics.

The casual reader of a statistic need not go through all of the questions. Just checking the first two questions, about the source

TABLE 7.1   Eight Questions to Ask about a Statistic

1.	What is the source of the statistic?
2.	Is there a methodology report?
3.	How were participants selected?
4.	What is the sample size?
5.	What is the response rate?
6.	What was done about missing data?
7.	How is crime defined and measured?
8.	How do the results compare with other information?

of the statistic and presence of a methodology report, will identify many of the unreliable statistics.

If you plan to quote a statistic—in a newspaper, magazine, journal article, term paper, fundraising appeal, or blog—the last six questions will help you ascertain whether the statistic is suitable for your purpose. The questions are not intended to substitute for professional statistical advice, but can help you evaluate the statistic and determine whether it fits your need.

Remember, all statistics have errors and even the best studies have unexpected glitches. We are not looking for perfection, but for an honest accounting that will allow you to evaluate how well the statistic answers your question.

WHAT IS THE SOURCE OF THE STATISTIC?

The first step is to find where the statistic originated. Was the statistic published by a government agency or in a reputable journal? This is not an infallible guide to quality—some articles in journals have errors and some statistics from other sources are carefully researched—but it does mean that someone with expertise in the area has reviewed the study.

Also look at who funded the study. Most medical journals require authors to disclose funding sources and conflicts of interest, and many other journals do as well. Of course, a conflict of interest does not necessarily mean the statistics are biased, but you may want to examine them more critically.

What to look for.   Find the original source of the statistic. If the source is not cited, and if you cannot find it through a search or by contacting the author, stop here. Do not trust the statistic.

Elder abuse study.   After an internet search, I found the AARP statistics on the webpage of the National Center on Elder Abuse, which in turn cited the original source, a 2014 article in the *Journal of General Internal Medicine* by Janey Peterson and colleagues. There are no apparent conflicts of interest for the authors or the organizations funding the study. So far, so good.

IS THERE A METHODOLOGY REPORT?

Scientific reports and government surveys always have a methodology or design section that explains how the study was conducted. This should have enough detail to allow someone else to replicate the study. The methodology report for a survey gives details about how the sample was selected, who responded, and what are the major sources of errors. It also contains a copy of the questionnaire or tells where it can be found.

The methodology report is written for specialists and is often lengthy and highly technical. For example, the methodology report for the Current Population Survey, which provides the official statistics on unemployment in the US, is 175 pages long—and is just one among hundreds of research reports.

All of the major national data sources discussed in Chapters 2 through 6 have methodology reports. The online supplement tells where you can find these.

What to look for.   At this stage, just ascertain that a methodology section or report exists. Do not trust a statistic if no one has described how the study was done.

You do not need to read the whole methodology report, but as we go through the questions I'll point out a few key features to look for.

Elder abuse study.   The Peterson article has a section titled "Design" that describes the survey design in detail. The population of interest was persons age 60 and older in the state of New York. Persons in nursing homes or other institutions, persons who spoke neither English nor Spanish, and persons who could not answer a short set of questions assessing cognitive ability were excluded.

The article is explicit about the study's limited scope: the results are about seniors in New York State who were eligible for the survey. No data were collected from other states, and you would need additional evidence—or to assume that the other states are similar to New York—to be able to say that the statistics apply to other locations.

HOW WERE PARTICIPANTS SELECTED?

With this question, you are looking for evidence that the researchers planned the study carefully and that results from study participants can be generalized to the population. Some crime statistics, such as the homicide statistics discussed in Chapter 2, are based on data from everyone in the population. But many other statistics come from samples, and we saw in Chapter 5 that the sample needs to be selected carefully for results to apply to the population.

Conveniently chosen samples. A statistic calculated from a conveniently chosen sample usually represents only the people who provided data to the sample—it cannot be taken to apply to the population at large.

Some news outlets have a "poll of the day" on their website or broadcast, where viewers can click, text, or call to volunteer their opinions about an issue. News outlets state that these polls are unscientific and do not represent public opinion, but sometimes other people quote the statistics without mentioning their unreliability.

For example, after the Cleveland Cavaliers had lost the first three games of the 2018 National Basketball Association championship series, a Phoenix television station asked in an online poll: "Can the Cavaliers come back and win the NBA championship?" When I checked the poll results at 2:40 pm, 9% of the respondents had answered "Yes, there's a chance" while the other 91% had said "No, it's over for Cleveland." The "yes" percentage increased to 12% after I participated (which, if you do the math, tells you that I was the 33rd person to respond to the poll). By recruiting others to take the survey, or by writing a computer program to flood the poll with responses, a determined Cleveland fan (or non-fan) could cause the poll to have any result desired.

Saying a sample is representative does not make it so. Although you are looking for evidence that the sample represents the population, the methodology report should give a detailed description of how the sample was selected, and not just state that "a representative sample was collected." Anyone can claim a sample is representative, but it is the method of data collection that makes a sample representative, not the use of the term. Convenient, easy-to-collect data are often unrepresentative.

Here is an example where the term "representative sample" may be misleading. According to the 2010 census, about 13% of US adults are between ages 18 and 24, 35% are between 25 and 44, 35% are between 45 and 64, and 17% are age 65 and over. A survey-taker might set up a web site where volunteers can click to take the survey, and stop collecting data from the respective age groups after obtaining responses from 130 persons age 18 to 24, 350 persons age 25 to 44, 350 persons age 45 to 64 and 170 persons age 65 and over. The sample has the same percentage in each age group as the census. But it still consists only of people who volunteered to participate, and their opinions and experiences on the questions of interest may differ from those of non-volunteers.

What to look for. Look for evidence that the data collected for the study can be generalized to the population of interest.

What kinds of sampling designs give representative samples? Some studies look at every record in a database (take a census), which gives a valid data collection if the database coincides with the population you want to study.

If the data are from a sample, look for the terms "probability sample," "stratified sample," "random selection," or "systematic sample." These terms indicate that the researchers made efforts to obtain a sample that was not just persons who volunteered to participate. If a probability sample was not used, look for an explanation of why not, and for evidence that the sample was not just the easiest-to-find members of the population.

Also look for evidence that the researchers took steps to ensure the quality of the data. For a survey, this includes testing the questions with prospective respondents to make sure they interpret them as intended, having training and continuing education for interviewers, and developing plans for what to do when people cannot be contacted or refuse to participate.

Elder abuse study. The Design section states that the data were collected through a telephone survey conducted between May and July, 2009. Telephone numbers to be called were selected using stratified random sampling from a database of residential telephone numbers—the term "stratified random sampling" implies that a probability sample was used. The survey oversampled nonwhite senior citizens and those age 70 and over.

Because only landline telephone numbers were called, persons living in households without a telephone, or with only a cell phone, were excluded from the study. In 2009, it is estimated that more than 90% of persons age 65 and over lived in a household with a landline. The article states that the results apply only to adults in households with landline telephones.

Peterson and her colleagues used professional survey interviewers and a questionnaire that had been extensively researched and tested. The interviewers received training on how to safeguard potential victims. Since they were asking about exploitation that may have been committed by family members, they also ascertained that the respondent was in a private place before administering the survey. To ensure that respondents understood the questions, the interviewers asked them to describe each incident of financial exploitation in their own words and the researchers used the narratives to classify incidents.

WHAT IS THE SAMPLE SIZE?

A television advertising trope dating to the 1950s asserted that "nine out of ten doctors agree" with a statement that followed. The implication is that 90% of all doctors agree with the statement. But of course the advertisement said nothing about how many doctors were asked or how they were chosen, and it could have been that only ten doctors were asked. This trope became so widespread that advertisers began making fun of it, showing a misfortune befalling the tenth doctor just as he was about to press the agree button.

A large sample size does not always mean that the survey is a good one—after all, internet polls of volunteers can get millions of responses but give misleading results—but a very small sample size usually indicates that a statistic is unreliable.

It is not just the total number of participants that matters here, but the sample size for each statistic that is computed. We saw in Chapter 4 that more than 160,000 persons were interviewed for the 2015 NCVS, but the data set contains only 103 incidents of rape or sexual assault. Any estimates about sexual assault victims—for example, what percentage reported the incident to the police, or what percentage knew their attacker—are calculated from those 103 incidents. The sample size for those statistics is 103, not 160,000.

A statistic computed from a small number of records has a large margin of error (MOE). The estimated percentage of sexual assaults reported to the police, from the NCVS, was 32.5% in 2015 and 22.9% in 2016. Thus, the reporting rate decreased by an estimated 9.6 percentage points from 2015 to 2016. But, because of the small sample sizes, the MOE for the change from 2015 to 2016 from was 19 percentage points—nearly twice as large as the decrease itself. The decrease can be explained by sampling variability—there is no evidence that the percentage of sexual assaults reported to the police actually decreased between 2015 and 2016.

Studies in which the entire population is measured usually do not report a MOE, but a small sample size can indicate that the statistic is unreliable. For example, the Supplementary Homicide Reports typically contain fewer than 15 incidents per year in which poison was used as the weapon. One can calculate that two-thirds of homicides by poison in 2016 were committed by women—but that statistic is based on 3 incidents with known male offenders and 6 with known female offenders.

What to look for.   The sample size (sometimes called n) is usually easy to find in the methodology report. Check that the study provides a sample size or margin of error for statistics calculated from subsets of the data.

Elder abuse study.   Interviews were obtained from 4,156 New Yorkers age 60 and over. This gives a relatively precise estimate of the prevalence of elder financial exploitation. However, only 195 of the respondents said they had experienced one of the five types of financial exploitation since turning 60, and all statistics about characteristics of victims—including the percentage who were victimized by a family member—are based on those 195 persons and have relatively large MOEs.

WHAT IS THE RESPONSE RATE?

The response rate, which is the percentage of persons selected for the survey who provided answers, provides some information about the quality of a survey, although it is not always a reliable measure of bias. Surveys with low response rates can give accurate results, and surveys with relatively high response rates can give biased statistics. It depends on the methods used to conduct the survey, on other factors such as the survey topic, and on whether the statistical methods used to adjust for nonresponse remove the potential biases.

That said, if all other things are equal it is generally better to have a high response rate than a low response rate. Censuses often have missing data too, and should report how much of the data is missing.

Survey response rates are decreasing. One challenge for almost all surveys is that response rates are much lower in 2018 than in the past. Figure 6.1 displayed this trend for the NCVS. Other large surveys that are conducted in person have had similar decreases in response rates.

A Pew Research report found that a typical response rate for a telephone survey conducted by an interviewer—considered to be the highest-quality type of telephone poll—decreased from 36% in 1997 to 9% in 2012 and 2016. Telephone surveys in which a computer reads a script, sometimes called "robopolls," are much cheaper to conduct than those employing interviewers, but they also have much lower response rates—typically less than 1%.

Low-response-rate probability samples vs. convenient samples. If only 1 out of every 100 persons selected for a probability sample participates, is the probability sample any better than one in which the sample consists of people who were conveniently available? After all, the respondents in a low-response-rate probability sample are those from whom it was easiest to collect data, and the survey fails to obtain responses from the majority of the persons selected to participate.

The advantages of probability sampling described in Chapter 5—that statistics from a sample can be applied to the population from which the sample is drawn and that the margin of error describes the accuracy of those statistics—assume that everyone responds. If there is nonresponse, you have to make assumptions about the nonrespondents. The lower the response rate, the more people you have to make assumptions about.

Probability samples that have nonresponse can sometimes yield statistics that are far from the population values. Some survey researchers argue that samples of volunteers are no worse than low-response rate probability samples, since the same weighting methods are used to adjust for potential bias in both types of samples and you need to assume those methods have taken care of the nonresponse bias.

But a probability sample, even one with a low response rate, has one big advantage. Although persons in the probability sample may refuse to participate, participation is limited to those who were randomly selected to be in the sample. The persons in a sample of volunteers can be anyone—a determined organization could arrange for people with a particular viewpoint to skew the survey, and participants might resemble the US population on race, sex, age, and every other characteristic except the question being asked. You may have seen this type of skewing with volunteered online reviews of books, movies, or restaurants.

The limited number of studies evaluating estimates from the two types of surveys have found that statistics from low-response-rate probability samples have less bias, when compared with results from high-quality surveys, than statistics from samples of volunteers.

What to look for.   First, is a response rate reported? A lack of a published response rate raises the suspicion that it may be low.

If reported, how is the response rate calculated? The survey should use one of the standard definitions for calculating response rates recommended by the American Association of Public Opinion Research, and should state which definition was used or give the formula.

It is usually better to have a high response rate than a low one, but there is no universal rule that tells when a response rate is too low for the statistic to be valid. It depends on how well the weight adjustment or imputation methods compensate for the missing data, which is the topic of the next question.

Elder abuse study.   The Peterson article stated that the response rate of 67% was calculated using an American Association of Public Opinion Research formula but did not give details of the calculations. A 67% response rate is high for a telephone survey from 2009. But nonresponse might still affect the statistics, and its effects should be investigated.

WHAT WAS DONE ABOUT MISSING DATA?

Almost all studies have missing data. In surveys, data are missing because some persons cannot be contacted or refuse to participate—nonresponse. In censuses such as the UCR, some agencies do not report data, or report for only part of the year.

Scope of the study.   Studies exclude some types of data, and should list these exclusions in the methodology report. The UCR has no information about crimes not reported to the police. The NCVS has no information about crimes against children, commercial establishments, or nursing home residents. The results of a study do not apply to persons out of its scope.

Differences between respondents and nonrespondents.   If persons who decline to participate in a survey have similar opinions and experiences to those who agree to participate, then results calculated from the respondents can apply to the population. The problem is that you cannot check whether respondents and nonrespondents have similar crime experiences, because nonrespondents do not provide that information.

But you may have some information about everyone selected to be in the sample. For example, a survey at a university may sample students from a database containing every student's age, sex, race, grade point average, and classes taken. You can compare these characteristics for the respondents and nonrespondents.

Even if you have no direct information about the nonrespondents in your sample, you can compare demographic and other statistics from your survey with those from a high-response-rate data collection such as the US decennial census or the Current Population Survey. If your survey estimates that 6% of the population is African American and the Census Bureau says that percentage is 13%, then you know that the nonrespondents are disproportionately African American.

Weighting and imputation.   If the study has a response rate less than 90%, check whether the researchers used weighting or imputation to adjust for missing data. Many surveys use a weighting method similar to that described for the NCVS in Chapter 6.

Do weighting and imputation reduce nonresponse bias? In most surveys they do, but you don't know for sure if they have removed the bias for a particular question unless you have another high-quality data source to compare with. The lower the response rate, the more the estimates depend on assumptions about the nonrespondents. The persons who do not respond to the survey could be like those who do respond on everything we know about them—except for their experiences with crime.

What to look for.   You do not need to get in the weeds of how the survey adjusted for nonresponse, but you can check for three things that will tell whether the study took missing data seriously.

1. Do the authors discuss limitations of the study?

2. Do they compare respondents and nonrespondents?

3. Is weighting or imputation employed to adjust for known differences between those who responded and those who did not?

Elder abuse study.   The elder abuse study clearly delineates its scope. The sample was drawn from New York residents who had landline telephones and could answer basic questions designed to assess cognitive ability.

The authors checked for differences between persons who consented to participate and those who refused. Age, region and household compositions were similar for the consenters and refusers, but persons who refused to participate were less likely to be married or partnered.

The article states that "weighted analyses were conducted for the primary outcomes," where the weights were calculated using US Census projections for age, race, and ethnicity composition in New York state.

But the published statistic estimating the lifetime prevalence of financial exploitation was calculated without the weights. The estimate of 4.7% was calculated as the 195 respondents who said they had been exploited divided by the total number of respondents, 4,156. The other statistics in the article were also calculated without weights, including those in the AARP quote at the beginning of this chapter. The article reported that 57.9% (113/195) of the perpetrators were family members, 16.9% (33/195) were friends or neighbors, and 14.9% (29/195) were home health aides.

The statistics published in the article accurately reflect the percentages for the 4,156 persons who responded to the survey. The respondents, however, were disproportionately female, African American, and over age 70, when compared with 2010 US Census data for New York state residents age 60 and over. If persons in those demographic groups are more (or less) likely to have experienced financial exploitation, then an analysis using nonresponse-adjusted weights would give a different estimate for lifetime prevalence, and the weighted statistic would be expected to be closer to the value for the population in New York.

HOW IS CRIME DEFINED AND MEASURED?

We have seen that crime statistics depend heavily on what types of events are counted and how questions are asked. Chapter 8 will discuss how different definitions and measurements of sexual assault lead to divergent statistics about its prevalence.

What to look for. Look for a clear definition of how crime is measured. For a survey, read the questions that were asked.

Elder abuse study. The survey asked respondents "Since you turned 60 years old has someone you live with or spend a lot of time with ever done any of the following" and then presented descriptions of types of financial exploitation: theft, coercion or fraud, pretending to be the respondent to obtain goods or services, and failure to make agreed-upon contributions to household expenses.

The descriptions gave explicit examples of each type of exploitation. For example, the first question asked whether anyone has "[s]tolen anything from you or used things that belonged to you but without your knowledge or permission? This could include money, bank ATM or credit cards, checks, personal property or documents." The fraud/coercion question asked about incidents in which someone "[f]orced, convinced, or misled you" to give property or rights such as "money, a bank account, a credit card, a deed to a house, personal property, or documents such as a will (last will/testament) or power of attorney."

The specific examples are likely to elicit more instances of exploitation than asking "Has anyone ever exploited you financially?" They also allow the researchers to distinguish among different types of exploitation. It is clear from the questions exactly what types of exploitation were measured by the survey.

The survey asked about incidents in which the exploitation was done by someone the respondent lived with or spent a lot of time with. It did not ask about financial exploitation or fraud done by strangers—for example, a stranger stealing the respondent's identity, or an "investment advisor" or fake charity soliciting funds over the telephone. Thus, although 113 out of the 195 respondents who reported financial exploitation said that they were exploited by a relative, those numbers do not include exploitations committed by persons that the respondent did not live with or spend a lot of time with.

In the AARP article cited at the beginning of the chapter, the statement that "[r]oughly 6 out of 10 cases of elder financial abuse are committed by relatives" was not qualified by mentioning that the study did not ask about financial abuse committed by strangers or casual acquaintances. If the survey had asked about exploitation committed by strangers, presumably the percentage of cases committed by relatives would have been smaller.

HOW DO THE RESULTS COMPARE WITH OTHER INFORMATION?

All of us evaluate new information by comparing it with what we already know (or think we know). If someone tells you that every day 15 women in the US are murdered by strangers (and you have not already deemed the statistic untrustworthy from one of the earlier questions), comparing this statistic with the CDC homicide statistics will give you reason to doubt it. Fifteen women murdered per day times 365 days gives 5,475 women murdered by strangers—much larger than the total number of murdered women in the 2016 CDC statistics, 3,895.

If the results of a study differ greatly from what is known from high-quality studies, you may want to explore why the results are different. Differences can often be ascribed to crime definitions, question wording, response rates, or procedures for collecting the data.

It is important to compare the results with high-quality studies, not just impressions about crime. Public opinion surveys taken periodically between 1993 and 2016 found that a majority of Americans thought that crime had increased since the previous year, even though the UCR and NCVS statistics both showed a large downward trend during the entire time period.

How would you critique the study if the numbers were different? If you enjoy drinking coffee at breakfast, and the morning newspaper reports on a new medical study saying coffee has health benefits, how do you evaluate that evidence? If you like coffee, there is a natural tendency to give more weight to a scientific article that says coffee is good for you than one that says coffee is bad for you. Flaws that might be ignored in the good-for-you study become reasons to doubt the conclusions of the bad-for-you study.

One way to counteract that tendency is to reread the study pretending that the results were different. Suppose the study said that coffee had health detriments instead of benefits. How would you critique it then? You may find upon rereading that the study participants were volunteers, or that it had a low response rate, or that it did not adjust for missing data. The study's findings of health benefits should then be interpreted in light of the new critique.

The same principle applies to interpreting crime statistics. If the estimated sexual assault rate from a survey had been higher than you expected instead of lower than you expected, how would you have critiqued the methodology of the study?

What to look for.   Ask:

1. How do the results compare with those of high-quality studies?

2. How would I critique the study if the results were different?

Elder abuse study.   The study was one of the first to study financial exploitation of senior citizens. The results of the study were consistent, however, with the limited information available on the subject. A supplemental survey conducted with the 2007 NCVS found that 4.2% of households headed by a person age 65 and older had been a victim of identity theft.

CONCLUSIONS FROM THE ELDER ABUSE STUDY

The elder abuse study was carefully planned and implemented, with the details clearly laid out in the methodology section of the article. It drew attention to the important problem of financial exploitation of senior citizens.

The article stated that 195 of the 4,156 respondents (4.7%) reported at least one financial exploitation event. That percentage and the other percentages in the article were calculated without using the sampling and nonresponse-adjusted weights. The statistics in the article describe the persons who participated in the study.

We do not know how accurately the percentages in the article estimate the prevalence of financial exploitation among New York residents age 60 and over (the population). In this survey, some race/ethnicity groups were oversampled and there was substantial nonresponse, causing the sample to have higher percentages of women and African Americans than the population.

As we discussed in Chapters 5 and 6, sampling weights compensate for the oversampling, and nonresponse adjustments to the weights often at least partially compensate for differences between the sample and population that result from nonresponse. Statistics calculated using the final weights would apply to the age 60+ population of New York. If women and African Americans tend to experience higher (or lower) levels of financial exploitation than other persons, then statistics calculated with the weights would differ from those in the article.

What about the AARP quote at the beginning of this chapter? The elder abuse study cannot answer the question of how many perpetrators of financial abuse in the US are known to their victims. The survey was conducted only in New York, which may be unlike other parts of the country.

More importantly, the survey asked only about financial exploitation done by persons well known to the respondent. It did not ask about financial exploitation or fraud committed by strangers or slight acquaintances. Thus, it cannot be concluded from this study that 90% of fraud perpetrators are known to their victims; the survey did not ask about frauds committed by persons not well known.

SUMMARY

There are no error-free statistics, but some statistics are more trustworthy than others. The eight questions in Table 7.1 can help you evaluate statistics you read.

Do not trust a statistic that is unsourced or from a study that does not describe how the data were collected.

If the statistic meets the tests in the first two questions, check to see what is actually measured by looking at who participated in the study (how were they selected, how many people participated, who was missing from the data and what was done about missing data), and how crime was measured.

All of us interpret statistics through the lens of our own experiences and opinions. One way to counteract prior expectations is to look at how you would critique the study if its conclusions were different.

CHAPTER 8

Sexual Assault

A 2016 STUDY BY THE US Government Accountability Office identified 10 separate data collections about sexual violence conducted by federal agencies since 2010. The estimates for 2011 "ranged from 244,190 rape or sexual assault victimizations to 1,929,000 victims of rape or attempted rape."

The estimate of 244,190 came from the 2011 National Crime Victimization Survey (NCVS). This was not the lowest estimate for the year: the FBI Uniform Crime Reports (UCR) recorded 84,175 rapes known to law enforcement agencies in 2011.

The estimate of 1,929,000 came from the 2011 National Intimate Partner and Sexual Violence Survey (NISVS). The Centers for Disease Control and Prevention (CDC) launched the NISVS in 2010 to study sexual violence, stalking, and intimate partner violence. The NISVS asks about past-year and lifetime experiences of sexual violence, and the health consequences of that violence. It is the source of the oft-quoted statistic that 1 in 5 women and 1 in 71 men have been raped at some time during their lives.

The NCVS and NISVS give very different estimates of the number of sexual assaults during 2011. And the difference is in the opposite direction than one would expect from the populations studied: the NISVS, interviewing men and women age 18 and older, would be expected to have a smaller number of assaults than the NCVS, which interviews persons age 12 and older.

What do these statistics actually measure, and how accurate are they? This chapter explores statistical properties of sexual assault estimates, and gives examples of how researchers have studied effects of measurement methods and missing data.

DEFINITIONS

When evaluating statistics about rape or sexual assault, first check how the terms are defined. Most definitions include three features: (1) a specific type of sexual act occurred, (2) the perpetrator used tactics such as physical force, threats of force, psychological coercion, or incapacitation, and (3) the victim did not provide consent for the act.

Until 2012, the UCR used the definition of rape that had been established in 1930: "the carnal knowledge of a female, forcibly and against her will." The definition implemented in 2013 expanded rape to include crimes against men and boys, other forms of sexual penetration, and offenses in which physical force was not involved: "The penetration, no matter how slight, of the vagina or anus with any body part or object, or oral penetration by a sex organ of another person, without the consent of the victim." For each year between 2013 and 2016, the UCR rape rate was more than 35% higher with the 2013 definition than with the 1930 definition.

The NCVS defines rape as "the unlawful penetration of a person against the will of the victim, with use or threatened use of force, or attempting such an act. Rape includes psychological coercion and physical force, and forced sexual intercourse means vaginal, anal, or oral penetration by the offender. Rape also includes incidents where penetration is from a foreign object (e.g., a bottle), victimizations against male and female victims, and both heterosexual and homosexual rape."

The NISVS rape definition emphasizes consent: "any completed or attempted unwanted vaginal (for women), oral, or anal penetration through the use of physical force (such as being pinned or held down, or by the use of violence) or threats to physically harm and includes times when the victim was drunk, high, drugged, or passed out and unable to consent."

Definitions of rape and sexual assault vary across surveys, and you often need to look at the survey questions to determine what types of unwanted sexual contact are included. In some studies (and in this book), rape is included in the category of sexual assault; for others (such as the NCVS), sexual assault consists of sexual violence other than rape.

LAW ENFORCEMENT STATISTICS

Measurement Errors

The UCR statistics on rape depend on how individual law enforcement agencies implement the definition when classifying incidents. The 2013 UCR manual states that the definition "includes instances in which the victim is incapable of giving consent because of temporary or permanent mental or physical incapacity (including due to the influence of drugs or alcohol) or because of age" but state laws vary with respect to what constitutes consent for a sexual act and age of consent.

Law enforcement agencies also vary with respect to the percentage of rape complaints that are not recorded or that are declared to be unfounded—"determined through investigation to be false or baseless." Unrecorded and unfounded complaints are not included in the UCR annual crime rate statistics.

In UCR data from 2009 to 2014, law enforcement agencies' percentages of rape complaints listed as unfounded ranged from 0% (which can occur either because the agency had no unfounded complaints, or because it did not report statistics about unfounded complaints to the UCR) to more than 50%.

Experts testifying in a 2010 US Senate hearing highlighted newspaper reports that five cities had systematically declared rape complaints to be unfounded, downgraded them to lesser offenses, or failed to create a written report. In one well-publicized case of Philadelphia during the 1990s, about one-third of sexual assaults had not been investigated and 2,300 cases had been erroneously downgraded or labeled as unfounded. The city subsequently implemented reforms and changed the systems for handling and classifying rape complaints.

The extent of this type of measurement error is unknown; the cases known have come to light because they were subjects of investigation. The departments studied in those investigations do not form a representative sample of police departments.

Missing Data

Crimes not reported to the police constitute the main source of missing data in the UCR. According to the NCVS, approximately 1 in 3 rapes and sexual assaults between 1993 and 2015 were reported to the police.

The reporting rate almost certainly varies across states and cities, although the NCVS does not have sufficient sample size in most localities to estimate it.

Social trends and norms can affect reporting rates. In 2017, the number of rapes reported in New York City spiked in October and November after having been on a downward trend the rest of the year. City officials speculated that part of the spike resulted from widespread news coverage in the fall of 2017 about public figures accused of sexual assault.

If one city has a higher UCR rape rate than another, it is difficult to say whether the city's actual rate is higher or whether a higher percentage of rapes are reported to the police. A city may have a higher UCR rate, for example, because victims feel more comfortable approaching its police department.

SURVEYS: MEASUREMENT OF SEXUAL ASSAULT

Many of the differences among estimates of sexual assault from surveys can be explained by the methods used to collect the data and measure sexual assault. Estimates are sensitive to the method used to collect the data (in-person interview, telephone interview, self-response through a mail or internet survey), to the recall period employed (are respondents asked about victimizations in the last 6 months? 12 months? lifetime?) and to missing data. Most of all, the statistics depend on how sexual assault is defined and how the questions are worded.

Before discussing how measurement methods affect the NCVS and NISVS, let's look at three studies that explored how question wording and other measurement decisions affect sexual assault estimates. These provide examples of how statistical designs and methods can be used to study measurement error.

The first study was a randomized experiment, the second compared estimates from two surveys conducted under similar circumstances, and the third examined what happened if estimates calculated from same set of respondents relied on different sets of questions.

Revision of the NCVS Questionnaire

NCVS respondents in the 1970s and 1980s were asked a series of screening questions such as "Did anyone beat you up, attack you or hit you with something, such as a rock or bottle?" None of the questions specifically mentioned rape or sexual assault—such terms were considered inappropriate for a government-sponsored survey in the 1970s. The interviewer recorded a rape only if the respondent volunteered that he or she had been raped.

The revised screening questions used since 1993 ask explicitly about rape. The revised assault question, "Has anyone attacked or threatened you in any of these ways," is followed by a list of specific types of attacks starting with "With any weapon, for instance, a gun or knife." The fifth item asks about "Any rape, attempted rape or other type of sexual attack."

The 1993 questionnaire added another specific cue: "Incidents involving forced or unwanted sexual activity are sometimes difficult to talk about ... Have you been forced or coerced to engage in unwanted sexual activity by (a) Someone you didn't know before, (b) A casual acquaintance, or (c) Someone you know well?"

As discussed in Chapter 6, half of the 1992 NCVS sample (chosen randomly) was administered the old questionnaire and the other half was administered the revised questionnaire. The half asked the old questions reported 0.7 rapes per 1,000 persons age 12 and over; the half asked the revised questions had 1.7 rapes per 1,000 persons. Changing the question wording increased the estimated rape counts and rates by nearly 150%.

Comparing Two Surveys of Women in College

Much of the research on measuring sexual assault has been conducted on women in college. They are a convenient population for university researchers to study, and colleges have lists of all students that make it easy to draw a sample. The disadvantage is that results do not necessarily generalize to women not in college.

Bonnie Fisher compared results from two telephone surveys that were intended to measure the same population: female students of 2-year or 4-year higher education institutions. The surveys were both conducted in 1997, had similar sampling designs, used interviewers from the same company who had the same training, and interviewed approximately the same number of women.

The major difference between the two surveys was the set of questions used to measure rape and sexual assault. The first survey used questions similar to those in the 1997 NCVS.

The second survey asked 12 behaviorally specific questions. The first question was "Since school began in fall 1996, has anyone made you have sexual intercourse by using force or threatening to harm you or someone close to you? Just so there is no mistake, by intercourse I mean putting a penis in your vagina." These questions did not ask the respondent to judge if she had been raped, but described the behavior that the survey classified as rape.

The survey with the behaviorally specific questions had much higher percentages of respondents who reported completed rape (1.66% vs. 0.16%) and attempted rape (1.10% vs. 0.18%) than the survey with the NCVS-type questions.

Although other differences between the surveys might have affected the results—for example, the introductory letter for the survey with the NCVS-type questions said the survey concerned "criminal victimizations" while the introduction for the other survey emphasized "unwanted sexual experience"—it seems likely that the behaviorally specific questions prompted more college women to report sexual victimizations to the interviewer.

Measurement Methods in an Internet Survey

The NCVS screening questions ask respondents whether any victimizations occurred in the previous 6 months. Persons answering yes to a screening question are asked follow-up questions about the details of each incident and asked to describe it in their own words. The screening questions indicate a victimization may have occurred; the follow-up questions and narrative are used to classify the type of crime (if any).

An NCVS respondent who did not mention rape in the screening questions will end up classified as a rape victim if his or her narrative of the crime describes a rape. Conversely, no rape will be recorded if the follow-up questions reveal that an incident mentioned in the screening questions took place more than 6 months ago or the details are not consistent with the definition of rape.

The NISVS uses the same set of questions to determine whether there were victimizations and to classify them. As an example, one NISVS question for women is: "When you were drunk, high, drugged, or passed out and unable to consent, how many people have ever had vaginal sex with you? By vaginal sex, we mean that a man or boy put his penis in your vagina." If a woman answers "one or more" to this or one of the other 8 questions describing acts that the survey defines as rape, she is classified as a rape victim.

Christopher Krebs and colleagues studied effects of using screening questions alone, as opposed to screening questions followed by additional questions about details, in a survey administered at 9 colleges in 2015.

A stratified random sample of undergraduate students was selected from each college. Sampled students were invited by e-mail to participate in an internet survey about sexual experiences and attitudes. Unlike the NCVS and NISVS, in which an interviewer asks questions, the students self-reported their answers on a computer, tablet, or smartphone.

Each respondent was asked a screening question about whether, and how many times, unwanted sexual contact had occurred since the beginning of the school year. The question defined unwanted sexual contact as "sexual contact that you did not consent to and that you did not want to happen. Remember that sexual contact includes touching of your sexual body parts, oral sex, anal sex, sexual intercourse, and penetration of your vagina or anus with a finger or object."

Those answering yes to the screening question were then asked about details of up to 3 incidents, including month of occurrence, type of sexual contact, and tactic used to perpetrate the contact (for example, physical force or incapacitation).

The research team could thus compare victimization rates calculated from the screener alone with rates calculated from the screener plus the additional information from the detailed reports. Both rates were calculated from the same set of 15,000 female respondents to the survey, so any differences could be attributed to the set of questions used.

The victimization rate calculated using the screener alone was 166 victimizations per 1,000 women; that from the screener plus detailed questions was 125 per 1,000 women. The detailed questions removed incidents that were undated, had not occurred during the school year, or did not have information on the type of unwanted sexual contact or the tactic used.

The study authors concluded that using the follow-up questions, as opposed to just the screening questions, produced more accurate estimates of victimizations occurring within a specific time period.

Measurement Differences for NCVS and NISVS

The NISVS has much higher estimates of rape and sexual assault than the NCVS. In 2010–2012, according to the NISVS, 1.2% of women age 18 and over had been raped in the previous 12 months— more than six times as large as the NCVS percentage of women age 12 and over who had been raped or sexually assaulted during each of those years. Can sampling variability and measurement methods explain this discrepancy?

Question wording.   Although the NCVS and NISVS have similar definitions of rape, they use different questions and approaches to measure it. The NISVS asks 9 (for women) or 11 (for men) behaviorally specific questions describing acts that are considered to be rape. The detailed questions list specific examples of non-consent, and ask about incidents that occurred when the respondent was drunk, high, drugged or passed out and unable to consent.

The NCVS screening questions are not as specific: two questions ask about rape and unwanted sexual activity (see page 101) but do not say what types of acts are included in those terms. The NCVS does not explicitly ask about incidents in which the victim could not provide consent because he or she was incapacitated.

Fisher's study of college women indicated that behaviorally specific questions elicit about ten times as many reports of rape as NCVS-type questions; if those patterns hold for men and for women not in college, then the question wordings may explain much of the difference between the NISVS and NCVS estimates.

Question context.   The NCVS is a criminal victimization survey, conducted under the auspices of the US Department of Justice. The screening questions about sexual assault are in the middle of a series of questions about crimes such as robbery, assault, and theft. Respondents know they are being asked about crime, and might not report incidents that they do not consider to be crimes.

The NISVS is presented to respondents as a public health survey conducted by the CDC. The questions about sexual violence are near the end of the survey. They follow questions about physical and mental health, psychological aggression by intimate partners (for example, has a romantic partner "ever told you that you were a loser, a failure, or not good enough"), coercive control and physical violence from intimate partners, and stalking.

Within the NISVS sexual violence section, the questions about rape come after questions about other unwanted contact such as "How many people have ever ... harassed you while you were in a public place in a way that made you feel unsafe?"

The context of the sexual violence questions, following questions about health, aggression by intimate partners, and harassment, may prompt NISVS respondents to think differently about their experiences and to report incidents of unwanted sexual contact that they do not consider to be crimes.

Memory of crimes.   Survey respondents often do not remember exactly when an event took place. The NCVS asks about crimes that occurred within the last 6 months; early research on the NCVS showed that persons asked about events in the last 6 months had more accurate recall than persons asked about events in the last 12 months. Some respondents may report incidents that occurred more than 6 months ago. The NCVS removes incidents that the person reported in an earlier interview (see page 66) and downweights incidents reported on a household's first interview.

After asking about lifetime experiences with rape and other unwanted sexual contact, the NISVS asks about how many of the events occurred within the past 12 months. It asks no follow-up questions about the dates of those incidents or other information that might help classify them more accurately. Krebs's study indicated that having follow-up questions in the NISVS would likely reduce the estimated percentage of persons who experienced rape in the previous 12 months.

NISVS estimates of the number of persons who experienced rape during the past year likely include some persons whose incidents should not have been counted in the statistics because they occurred outside of the time window or were misclassified.

Both surveys fail to capture incidents that a respondent does not recall or is unwilling to discuss with an interviewer.

How the survey is administered.   Differences in interviewing procedures may affect responses to the survey.

The NISVS contacts persons by telephone and interviews them once. Landline and cell phone numbers are sampled, but persons without telephones are excluded.

Each person in the NCVS is interviewed up to seven times at 6-month intervals. The first NCVS interview with a household is conducted in person; subsequent interviews are either in person or by telephone. The respondent may develop a rapport with an NCVS interviewer over repeated contacts, or familiarity with the survey may lead to different answers on later interviews than earlier interviews.

Both surveys attempt to interview people in private. As mentioned in Chapter 6, sometimes other people are present in the room during in-person NCVS interviews and there is concern that respondents may report fewer victimizations when others are present. It is difficult to tell whether anyone else is listening during a telephone interview.

Both the NISVS and NCVS employ interviewers to ask the survey questions. In other surveys (such as the Krebs study), respondents answer the survey questions directly on paper or on a computer. Self-administered surveys avoid interviewer effects and may afford more respondent privacy, but also lack an interviewer who can clarify questions or help with a narrative description.

Sampling variability.   The NCVS's large sample size (143,000 persons in 2011) results in small margins of error (MOE) for annual victimization rate estimates. The MOE for the sexual assault victimization rate, which counts multiple incidents per person, is typically less than 0.5 victimizations per 1,000 persons. For estimates of the percentage of persons victimized by rape or sexual assault in the previous year, the MOE is even smaller—less than two hundredths of a percentage point.

The NISVS interviews about 10,000 to 16,000 persons each year. Its estimates thus have much larger MOEs. For 2011, the NISVS MOE for the percentage of women victimized by rape was about one half of a percentage point.

Remember, though, that the MOE does not include uncertainty about estimates arising from measurement error or nonresponse.

Conclusions.   With all of the differences between the NCVS and NISVS, it would be strange if they did *not* have different estimates.

The two surveys present different pictures of sexual violence in the US. The NISVS, with its more expansive concepts of sexual violence, measures crimes and forms of sexual misconduct that are not typically classified as crimes. So far, the NISVS has too few years of data to be able to tell if the two surveys show similar trends in sexual violence over time.

The NISVS produces much higher estimates of how many women experienced rape in the previous 12 months than the NCVS, but the survey questions show that the two surveys are measuring different concepts. The behaviorally specific questions of the NISVS, and the context as a public health survey studying many forms of sexual misconduct, likely elicit more reports of incidents than the NCVS questions. The NISVS estimates are also higher because they do not exclude some of the incidents that would be determined to be out of scope if follow-up questions were asked.

And there is one more difference between the surveys that affects the accuracy of the statistics: the amount of nonresponse.

SURVEYS: NONRESPONSE

The NCVS and NISVS both have nonresponse, but the NISVS has much more of it. In 2010 and 2011 the overall response rate for persons in the NCVS was approximately 80% (see Figure 6.1). The NISVS response rate was about 30%.

Both surveys adjust the weights of the respondents to try to reduce nonresponse bias. The NCVS weight adjustments, described on page 75, distribute the sampling weights of nonrespondents to respondents that have similar demographic characteristics. The NISVS uses similar procedures.

Although the weight adjustments are thought to reduce nonresponse bias, they likely do not eliminate it. How can you learn about nonresponse bias when you do not know the nonrespondents' experiences? The next section describes an experiment that investigated possible effects of nonresponse.

Nonresponse in a Survey on Sexual Assault in Universities

David Cantor and colleagues explored effects of nonresponse in another internet survey about sexual assault and misconduct, administered at 27 universities in April and May of 2015.

E-mail requests with a link to the survey were sent to all enrolled undergraduate, graduate, and professional school students age 18 or over at each university. Students who did not respond to the first survey invitation were sent up to two reminder e-mails.

Overall, 19% of the students who were asked to participate in the survey did so. Even though the weights were adjusted for nonresponse using information about the student populations from university databases, such a low response rate raises concerns about potential nonresponse bias.

Comparing respondents in different incentive groups. The nonrespondents' sexual assault experiences were unknown.

But the researchers had conducted a randomized experiment within the survey. Many surveys offer people a small amount of money as an incentive to participate. In each of 19 of the schools, 6,000 students were randomly placed in the "gift card" group and told they would receive a $5 gift card upon completing the survey. The remaining students in each school, the "alternative" group, were offered a less valuable incentive such as being entered in a drawing for a $500 cash prize.

The response rate in the gift card group was 26%—9 percentage points higher than the response rate of 17% in the alternative group. It appeared that the promise of a $5 gift card persuaded some students to respond who otherwise would not have.

Were the extra students who responded because of the promised gift card more, less, or equally likely to have reported being victims of sexual assault? If more likely, then the gift card group would have higher estimated sexual assault rates than the alternative group; the incentive may have persuaded students to participate who otherwise would not have wanted to share their experiences. If less likely, the gift card group would have lower estimated sexual assault rates than the other group, and the incentive may have persuaded non-victims to make the effort to take the survey.

The gift card group had about the same or significantly lower estimated rates than the alternative group for each of the five types of sexual assault or misconduct measured in the survey. The students who were less likely to respond to the survey (who would not respond without the extra incentive) were also less likely to have reported victimizations. If the nonrespondents are similar to the students who need the extra incentive, then the estimates of sexual victimization from the survey are too high.

Comparing early and late respondents.   The researchers also compared students who took the survey before the first reminder e-mail (early respondents) with those who responded after being reminded (late respondents).

Late respondents had significantly lower rates of four types of sexual victimization (penetration by physical force or incapacitation, sexual touching by physical force or incapacitation, nonconsensual sexual contact by absence of affirmative consent, and sexual harassment) than early respondents.

If late, more reluctant, respondents are more similar to nonrespondents, then differences in the sexual victimization rates between early and late respondents indicate that nonresponse bias remains after the weight adjustments.

Conclusions. Both investigations indicated that the harder-to-get respondents—those who were incentivized by the promise of a gift card, or those who responded later in the data collection period—had lower rates of sexual assault than the easier-to-get respondents. If the nonrespondents are like the harder-to-get respondents, then the survey's estimates of sexual assault rates are too high.

This is just one study, and does not imply that other surveys with low response rates have bias in the same direction. But it does imply that response rate is important for surveys of sexual assault, and that surveys should try to have as few nonrespondents as possible and perform thorough analyses to explore potential bias from nonresponse. Instead of e-mailing every student at a university, it would be better to take a smaller sample but use more of the survey budget to try to increase the response rate.

SEXUAL ASSAULTS AGAINST CHILDREN

Sexual assaults against children are undercounted in every data source. The NCVS does not interview children under the age of 12 and has no information about assaults against children.

The NISVS interviews persons age 18 and over and likewise cannot produce estimates of the percentage of children assaulted in the past year. But the NISVS asks respondents about their lifetime experiences with sexual violence and when they were first victimized, and thus provides some information on how many adults were sexually assaulted as children. Of the persons who reported having been raped in their lifetime, the first rape had occurred before age 11 for an estimated 11% of females and 28% of males.

Some surveys have asked young children's parents or guardians about sexual abuse their children have experienced, but the studies' sample sizes for young children have been too small to produce reliable estimates. It is also likely that the parents or guardians would not report abuse that they initiated or tolerated.

The largest data sets on sexual abuse of children come from law enforcement and child protective services agencies. These data, of course, are limited to cases known to the agencies, and we do not know what percentage of children's victimizations are reported. But the agency data provide lower bounds for the number of sexual victimizations of children; they also provide an independent data source for investigating measurement errors in survey data.

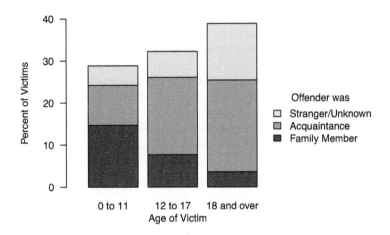

FIGURE 8.1 Sexual assaults by victim age group, from 2015 NIBRS.

The US National Child Abuse and Neglect Data System, which collects and analyzes state-submitted data on child abuse and neglect, reported that 683,000 children had been determined by child protective services agencies to have been abused or neglected in 2015. In 97% of all child maltreatment cases, at least one of the perpetrators had been a parent, guardian, or relative. About 57,000 of the cases involved sexual abuse.

The 2015 UCR National Incident-Based Reporting System (NIBRS) data contained about 77,000 incidents of rape, sodomy, sexual assault with an object, and fondling where the victim's age was known. Figure 8.1 shows the age distribution of the victims of these offenses and the percentages perpetrated by a family member, acquaintance, or stranger or unknown person.

Even though only 1/3 of law enforcement agencies reported their data to NIBRS in 2015 and the statistics are not nationally representative, the fact that about 22,000 (30%) of the incidents had a victim under age 12 shows that surveys excluding children substantially underestimate the amount of sexual assault in the US. Family members were responsible for more than half of the sexual assaults against children under age 12, implying that surveys of parents also likely underestimate these.

SUMMARY

Rape and sexual assault are among the most challenging crimes to measure. Statistics from law enforcement agencies capture only assaults that are reported to and recorded by the police. Statistics from surveys are highly sensitive to the question wording and to nonresponse; in addition, some victims may not report incidents, or may report incidents that are outside of the survey's scope.

Crimes against children, and other populations typically excluded from surveys such as homeless persons, are undercounted in all data collections.

Randomized experiments and statistical analyses can identify some of the effects of measurement error and nonresponse. Survey questions need to be extensively tested to ensure that they are measuring the concepts of interest. Having multiple sources of data provides a more comprehensive picture of sexual assault than any source on its own, and information in one source can provide insight into measurement error in another source.

CHAPTER 9

Fraud and Identity Theft

W HEN BERNARD MADOFF was arrested on December
11, 2008 for securities fraud, he confessed that the fraud
could have topped $50 billion in what is thought to be the largest
private Ponzi scheme in history. Prosecutors later estimated the
size of the scheme as $64.8 billion.

Madoff promised investors high regular returns regardless of
economic conditions, but in fact he had created fake trading reports
and paid early investors with funds provided by later investors. He
had also siphoned off large amounts of investors' funds for himself.

If you had asked Madoff's investors in 2005 whether they were
victims of fraud, however, it is likely that most would have said
no. They were unaware that they had been defrauded. Although
most of the victims who later submitted impact statements to the
court said they had never met Madoff, they had great trust in him
and did not consider that they had put themselves in a vulnerable
financial situation. They learned they had been victims of fraud
only after Madoff was arrested.

MEASUREMENT CHALLENGES

Of all the crimes discussed in this book, fraud and identity theft
may be the most difficult to measure. Each data source includes
only a portion of the crimes that have occurred, and many crimes
are undetected.

Statistics from law enforcement and other government agencies are limited to frauds that they uncover or that are reported to them, thought to be a small fraction of the total.

Surveys fail to capture incidents in which the victims are unaware they have been defrauded. And of course all survey estimates depend on how respondents interpret an event.

Persons who know they were victims of fraud are often unwilling to report the crime to a police department, government agency, or survey. Senior citizens may fear that family members who learn about their victimization may decide they are no longer competent to handle their own affairs. Persons victimized by a relative may be reluctant or afraid to report the incident. Other victims may be embarrassed or ashamed to say they fell for a scam.

The issues are similar for frauds against organizations. A business or charitable organization may fear that reporting a fraud might lead to bad publicity and diminished public trust.

As with all other crimes, estimates depend on what types of offenses are counted.

DEFINITIONS AND CLASSIFICATION

Fraud. According to the US Department of Justice, "Fraud occurs when a person or business intentionally deceives another with promises of goods, services, or financial benefits that do not exist, were never intended to be provided, or were misrepresented." Examples include:

Telemarketing fraud. A solicitation offering phony or misrepresented goods or services.

Credit card fraud. Unauthorized use of a credit or debit card number to make fraudulent purchases.

Investment fraud. Using false claims to solicit investments, as with Madoff's Ponzi scheme.

Charity fraud. Collecting money for a fake or misrepresented charity, religious organization, or educational institute.

Advance fee schemes. The victim pays money in anticipation of receiving something of greater value, but in fact receives little or nothing. An example is the e-mail scam in which a foreign "prince" offers a percentage if you will help transfer a large sum of money—after you demonstrate your willingness to help by sending an advance fee.

Affinity fraud. An investment scam that targets members of an identifiable group such as a church, social organization, or ethnic community. Often the fraudster is, or pretends to be, a member of the group.

Romance scams. A fraudster pretends to be romantically interested in order to obtain money.

Technical support fraud. A fraudster, impersonating a technical support representative, asks victims to send money or provide remote access to their computers.

Government impersonation schemes. A fraudster pretends to be from a government agency and demands payment of "fines" for underpayment of taxes, failure to appear for jury duty, or another fake claim.

For an incident to be considered fraud, the fraudster must intentionally deceive the victim into participating in the transaction. As with theft, the victim loses something of financial value. In a theft, however, the thief takes something without the victim's consent. The fraudster persuades the victim, through deceit or lying, to provide money, goods, or services.

Identity theft. In **identity theft**, a thief assumes someone else's identity in order to commit a criminal act. The theft of the information occurs without the victim's consent. The thief may then, for example, use the victim's checking account or credit cards, open new financial accounts using the victim's identity, file a fraudulent tax return in order to receive the victim's tax refund, or purchase weapons under the victim's name that are later used in a robbery.

Identity theft often occurs together with fraud. A criminal may use someone else's health insurance account to obtain medical care. The owner of the health insurance account is a victim of identity theft; the clinic providing the care is a victim of fraud.

New types of fraud and identity theft are continually being created, which complicates attempts to categorize them. For example, in a traditional identity theft, a criminal might open new credit cards using the victim's true identity. In a variant called synthetic identity fraud, a criminal creates fake identities by combining real social security numbers with fictitious names and addresses.

Classification. Agencies and surveys reporting fraud statistics use different definitions, terminology, and classification schemes. I counted more than 80 terms that US government agencies use to categorize types of fraud and identity theft, including terms related to:

- Victim selection (e.g. fraud against the elderly, affinity fraud)

- Perpetration techniques (e.g. impersonation, advance fee schemes)

- Contact method (e.g. mail, telephone, or e-mail solicitation)

- Payment method (e.g. wire transfer, credit card)

- Promised benefit to the victim (e.g. business opportunity, lottery or sweepstakes, home repair)

A fraud scheme might be classified according to any of these dimensions: someone impersonating a member of the armed services (perpetration technique) might ask recently divorced women (victim selection) to wire funds (payment method) after deceptively promising a romantic relationship (promised benefit) through a series of e-mails (contact method).

A taxonomy developed in 2015 for the National Crime Victimization Survey (NCVS) classifies frauds as belonging to exactly one of seven types of promised benefits. Additional variables may be used to classify the incidents by other dimensions.

STATISTICS FROM GOVERNMENT AGENCIES

The online supplement lists more than 20 US federal agencies that investigate and collect statistics about fraud and identity theft. Here are three of the data collections:

Uniform Crime Reports (UCR). Fraud and identity theft are not tallied in the Summary Reporting System of the UCR (see Table 3.1), but the National Incident-Based Reporting System (NIBRS) collects statistics about frauds and identity thefts reported to the police. In 2016, the states and agencies participating in NIBRS (about 37% of law enforcement agencies) reported a total of 252,000 fraud and identity theft offenses.

The NIBRS statistics on fraud are far lower, even accounting for the incomplete participation, than statistics from other sources.

Many frauds are reported to another agency instead of to a police department, or not reported to any authority.

Other FBI Statistics. In 2017, the FBI Internet Crime Complaint Center received more than 300,000 complaints with more than $1.4 billion in associated losses. The three crimes generating the highest financial losses, out of the 33 types reported, were business e-mail compromises (e-mail accounts were compromised to conduct unauthorized fund transfers), romance frauds, and non-payment (goods are delivered but not paid for) and non-delivery (goods are paid for but not delivered) schemes. Data are not publicly available, but the annual reports provide statistical summaries by type of crime, victim age, and state.

Other divisions of the FBI also collect statistics about fraud from investigations of criminal activities such as money laundering, securities fraud, mortgage fraud, mass marketing fraud, embezzlement, and cybercrimes.

Federal Trade Commission (FTC). The FTC Consumer Sentinel Network contains data from consumer complaints of identity theft, fake sweepstakes, advance fee schemes, and other frauds. The FTC collects reports directly from consumers, state Attorney General offices, other federal agencies such as the US Department of Defense and the Postal Inspection Service, and private-sector organizations such as Better Business Bureaus and financial services companies.

Although the data are available only to law enforcement agencies, the FTC publishes annual statistical reports about the complaints received. During 2017, the FTC received nearly 1.1 million fraud complaints; about one-fifth reported a loss, totaling $905 million. One of the most common types of fraud, with 350,000 complaints and $328 million reported lost, was imposter scams, in which someone poses as a government representative, relative, technical support agent, or reputable business. The FTC also received 371,000 reports of identity theft in 2017.

Measurement Error and Missing Data

Agency data collections use different definitions and classification systems for fraud, making it difficult to compare statistics.

NIBRS reports crime counts for categories including false pretenses, swindling, bad checks, credit card fraud, impersonation, wire fraud, and identity theft. The FBI Internet Crime Complaint Center uses different categories including data breaches, romance schemes, extortion, and lottery/sweepstakes schemes. The FTC reports use yet another classification scheme with 30 types of fraud and 7 types of identity theft.

All agency statistics have missing data from the large number of frauds and identity thefts that are unreported. Because of this, all of the considerations from Chapter 3 about interpreting statistics from law enforcement agencies apply to statistics about fraud.

The annual fraud statistics from government agencies always underestimate the true amount of fraud, and a perceived increase in fraud may occur because more people are reporting it, not because there has been a real increase. Changes in statistics can reflect changes in enforcement efforts rather than changes in the true amount of crime.

STATISTICS FROM SURVEYS

The NCVS, when initiated in the early 1970s, focused mainly on the same set of crimes as the UCR, and did not measure fraud or identity theft.

Since 2004, however, a supplementary questionnaire on identity theft has been administered (in some years) to NCVS respondents. In 2014 an estimated 17.6 million persons were victimized by identity theft, with total financial losses of $15 billion.

The estimated total financial loss from the robberies, burglaries, and thefts in the 2014 NCVS was about $12 billion for the 15.9 million victimizations. Thus, according to the NCVS, there were more victims and higher financial losses just from identity theft than all of the "street crimes" combined.

The identity theft supplement does not ask about other types of fraud. Data were collected for the first NCVS Supplemental Fraud Survey in 2017, with estimates to be published in 2019.

There have been other surveys about fraud, but most have been much smaller than the NCVS, with much lower response rates.

The FTC has conducted three telephone surveys about fraud. The 2011 survey asked 3,638 respondents (response rate 14%) about the types of fraud appearing most frequently in the FTC's Consumer Sentinel database. It estimated that about 11% of adult Americans (26 million persons) had been victimized by fraud in 2011.

Measurement Error and Missing Data

All of the issues of measurement error and missing data, discussed for the NCVS in Chapter 6, apply to surveys about fraud and identity theft.

One limitation of surveys is particularly pertinent for estimating the amount of fraud and identity theft. The NCVS and most other surveys do not interview persons in institutions such as nursing homes, nor do they interview persons with Alzheimer's disease or other cognitive impairments. These populations are thought to be particularly vulnerable to crimes of fraud, and statistics about them must be obtained from non-survey sources such as enforcement activities. An investigation by *USA Today* found that between 2010 and 2013, government inspectors had issued more than 1,500 citations to nursing homes for mismanaging residents' trust funds or failing to protect them from theft.

The NCVS also does not survey members of the military living in barracks or abroad, who are frequent targets of fraudsters. In 2017, the FTC Consumer Sentinel Network received more than 80,000 fraud and identity theft reports from the military community.

There have not been as many studies about measurement error for surveys about fraud as for surveys measuring sexual assault. The limited studies that have been done, however, indicate that survey question wording has a large effect on fraud statistics. Questions need to elicit reports of fraud (which the respondent may be reluctant to discuss), but also be able to distinguish between a fraud and a negative consumer experience.

Some respondents may report incidents that were not, in fact, fraud. Fraud involves intentional deception, and a person responding to a survey (or making a complaint to the FTC) may not know the intent of the presumed offender. Was the investment opportunity fraudulent, or did the business venture just fail? A person may think he was a victim of fraud when in reality he simply received what he perceived to be poor service. It is not fraud when a restaurant serves an overcooked hamburger.

Statistics from surveys depend on how accurately the respondents report their fraud and identity theft experiences. If a large number of respondents do not believe they were fraud victims, or are unwilling to report the victimizations to the survey, then the statistics from the survey will likely underestimate the amount of crime and financial losses.

Victims Who Do Not Report Fraud to a Survey

How many fraud victims continue sending money without realizing they have been defrauded? Or how many will not mention the incidents on a survey? A survey asking people about their experiences with fraud victimization will not be able to answer that question—unless it is known from another data source that the respondents have been fraud victims.

In 2011, the AARP Foundation surveyed persons who had been listed as victims in fraud schemes that had been discovered by law enforcement agencies. Five of the schemes had involved investment fraud, offering opportunities to invest in nonexistent oil wells, movie productions, gold, or commercial storage facilities. Three other schemes had involved fake business opportunities, prescription drug discounts, and protection from identity theft. The remainder had asked victims to send an advance fee to receive a loan or collect prize money from a lottery. All persons on the lists had been independently verified to have been defrauded.

The researchers asked persons on the lists to participate in a telephone survey about consumer opinions and experiences. Importantly, they did not say the survey was about fraud, which could have influenced the answers. The survey asked about computer use, hypothetical interest in different types of sales promotions such as an online discount prescription program, behavior such as how often they entered drawings to win a free gift, general knowledge of consumer rights, and recent life events such as change in employment status or divorce.

After all of these general questions, respondents were asked whether they had lost money in a financial transaction, or had been misled or defrauded.

Of the 723 respondents—all known fraud victims—56% of persons under age 55 said they had been defrauded. Among the persons age 55 and older, only 37% said they had been defrauded.

This study is an example of valuable information that was obtained from a conveniently chosen sample. The sample is not representative of the population of fraud victims as a whole—it was limited to victims from a handful of schemes that had been uncovered by law enforcement agencies. But the low percentage of persons who said they had been defrauded implies that any victimization survey that relies on self-reports of fraud likely underestimates the amount.

MULTIPLE DATA SOURCES

Each data source on fraud and identity theft captures only part of the incidents.

Data sets from the FBI, the FTC, and other government agencies contain only incidents that are reported to them. How much do they miss?

The 2014 NCVS identity theft supplement estimated that only 8% of victims said they had reported the identity theft to the police, and only 1% to the FTC. Some of these incidents may have become known to law enforcement or the FTC through other channels, but these percentages indicate that a large number of identity thefts do not end up in the NIBRS or FTC data.

The AARP study surveying known fraud victims indicates that a large number of frauds are also not reported in surveys. In addition, surveys often exclude vulnerable groups such as persons in nursing homes.

One approach that could be taken to obtain better estimates of fraud prevalence would be to combine information from the different sources. This would likely still miss many crimes, but prevalence statistics from the combined data would be larger (and presumably, closer to the true values) and provide more information on the strengths and deficiencies of the individual data sources.

Figure 9.1 depicts three types of data sources for fraud. The circle sizes do not indicate the numbers of crimes captured by each source; I drew them with equal sizes only because I do not know how many crimes are in the different data sources and their overlaps.

To estimate the total amount of fraud in the combined three circles of Figure 9.1, we must be able to estimate how much fraud is in the overlapping parts of the circles, measured by more than one source. If we simply add the estimates from the three separate circles, the crimes in the overlapping parts of the circles will be double- or triple-counted.

The first two sources, surveys and agency data, contain some of the same fraud incidents. The 2017 NCVS Supplemental Fraud Survey will provide estimates of how many fraud incidents from the survey were reported to the police, the FBI, the FTC, or another agency. After the data from the survey are available, we will be able to estimate how many fraud victims are in the overlap of the NCVS circle and the agency data circle.

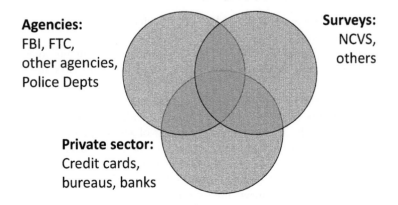

Agencies:
FBI, FTC,
other agencies,
Police Depts

Surveys:
NCVS,
others

Private sector:
Credit cards,
bureaus, banks

FIGURE 9.1 Data sources on fraud and identity theft

Surveys and agency data also contain information about incidents not known from the other source. The survey data allow estimation of the number of incidents not reported to any agency. The agency data include frauds that have been uncovered by law enforcement but not reported to the NCVS, such as those against nursing home residents and persons unaware or unwilling to say they have been defrauded.

The private sector provides another source of data. Many frauds and identity thefts are detected by credit card companies and financial institutions. According to the 2014 NCVS, about 87% of identity theft victims reported the thefts to a credit card company or bank; 8% reported them to a credit bureau.

Financial services associations regularly perform investigations and take surveys on fraud affecting their institutions. For example, the American Banking Association tracks deposit account fraud and takes surveys to estimate fraud losses as well as the dollar amounts of fraud attempts that were stopped by banks. Organizations that incur losses from fraud may have some of the best records about it.

Each of the circles in Figure 9.1 itself consists of multiple data sources. Combining estimates from different sources would require intensive research, particularly to estimate how much fraud is in the overlapping parts of the circles. Of course, the crimes not captured by any circle would still be missed.

SUMMARY

All of the statistics discussed in this chapter underestimate the extent of fraud and identity theft.

Even with that underestimation, however, these crimes appear to affect more Americans, and result in higher financial losses, than all robberies and traditional property crimes such as burglary, theft, and motor vehicle theft put together.

Multiple sources of data are needed to measure fraud and identity theft. Victims who report their crimes can report to a number of different agencies, and as of 2018 there have been few attempts to consolidate and reconcile the various statistics.

The NCVS started measuring fraud in 2017, and that will provide a badly needed source of information on crimes that are not reported to authorities. The agency data, by contrast, can provide information about frauds that victims are unaware of or unwilling to report.

Even with multiple sources of information, fraud estimates will be too small. But the different sources can provide information about measurement errors and missing data in the other sources, as well as providing additional information about types of fraud that can be used by enforcement and investigative agencies.

CHAPTER **10**

Big Data and Crime Statistics

D ATA ARE THE NEW OIL. So says *The Economist* and the companies that extract, refine, manage, and sell data. When you use social media, make a purchase with a credit card, travel accompanied by your cell phone, click on a web site, or walk down a city street, someone is capturing information about you.

This chapter explores the potential of "big data," the massive data collections that accompany the internet age, to investigate aspects of crime that are less easily studied from other sources.

Police departments too use big data to allocate resources, guide crime-deterrence programs, and inform police responses to crime complaints. How does data-driven policing affect crime statistics?

LAW-ENFORCEMENT-RELATED DEATHS

According to the Uniform Crime Reports (UCR), each year between 2012 and 2016 had about 450 justifiable homicides. The CDC statistics from death certificates reported about 500 legal interventions (deaths due to law enforcement actions) for each year in the same time period. Do these numbers capture all of the deaths caused by law enforcement officers?

The Guardian newspaper asked readers to send information about fatalities resulting from encounters with law enforcement personnel in 2015 and 2016. They supplemented this information with information from their own reporting, web searches of news reports, and information from other websites that tracked deaths from law enforcement officers. *The Guardian* reported that 1,146 deaths had resulted from encounters with law enforcement in 2015 and 1,093 in 2016—more than twice as large as the FBI and CDC statistics.

Since anyone can post a report on the internet or send information to a newspaper, it is important to verify data from these sources. *The Guardian* confirmed with law enforcement agencies and medical examiner's offices that persons in their database had been killed as a result of law enforcement activities.

Did *The Guardian* capture all of the deaths? You can sometimes estimate the total number of people in a population if you have partial lists of the population from two or more organizations. Merging the lists will create a longer list consisting of persons on at least one of the partial lists. Moreover, by looking at the numbers of people on one list but not others, you can, under certain assumptions, estimate how many people in the population are missing from all the lists. If there is a lot of overlap among the lists, then that estimated number will be small.

Justin Feldman and colleagues compared *The Guardian's* list with the CDC's list of legal intervention deaths. They found that 1,086 records in *The Guardian's* 2015 data set met the CDC's definition of legal intervention, and requested the cause-of-death code for those records from the CDC's 2015 database of death record information.

Matching the records in *The Guardian's* database with CDC records is more challenging than it might seem. The same person can be known by different names (Michael, Mike, Mickey), different persons can have the same name, and information on victim age or date of death can be wrong. A computer algorithm evaluates similarity between each record from *The Guardian's* data and records in the CDC database, and returns CDC records with high similarity scores as possible matches. The researchers used date and state of death to verify which of the possible matches was likely to be the person in *The Guardian's* database, and identified matches for 991 of the 1,086 names.

The CDC had classified 444 (44.8%) of those 991 records as legal interventions and the remaining 547 in a different category (mostly assault, which meant the deaths were counted as homicides). Assuming that the 95 unmatched records and the records not in *The Guardian* database have the same percentage (44.8%) with a CDC legal intervention classification, the researchers estimated a total of 1,166 law-enforcement-related deaths in 2015 (see the online supplement for details). *The Guardian's* crowdsourcing and web searches had captured almost all of the estimated total number of deaths.

Evaluating Statistics from Big Data.   Statistics from big data need to be evaluated using the same statistical principles as for other statistics. What is the likely size of the difference between the statistic and the true value? How do missing data and measurement error affect the statistics?

The Guardian's investigation relied on reader and online media reports to identify cases that were not in the official statistics. Investigations by the Bureau of Justice Statistics and *The Washington Post*, which also relied on media reports, obtained similar estimates of deaths caused by law enforcement. These studies provide corroboration that the UCR and CDC statistics through 2016 missed at least half of the deaths.

One concern that we often have about convenient data—that some of it might be fictional—does not apply here. Each investigation independently verified that the deaths identified from the media sources were indeed related to law enforcement. This reduced the amount of measurement error in the estimates.

The main source of potential error is that each estimate may still be missing some of the law-enforcement related deaths. Legal intervention deaths that are misclassified by the CDC may also be less likely to be reported by news media. But the statistics that rely on multiple sources are likely to be much closer to the truth than the UCR and CDC statistics by themselves.

Web pages and media reports may have information about high-profile events such as police-involved deaths, but they are less likely to have information about less newsworthy events. The value of web information diminishes with time because information can change or be purged, or the web pages may be taken down. Web-based data are fragile and mutable.

Although *The Guardian's* approach worked well for estimating law-enforcement-related deaths, such an approach would probably not work as well for estimating events that are less publicized.

FRAUD

Many people first learn they were victims of fraud when notified by their credit card companies. Credit card companies use algorithms to detect possibly fraudulent transactions among the huge number of transactions processed every day.

Your credit card transactions tend to follow patterns, and the algorithms look for deviations from these patterns. For example, you live in Oregon but your card number is used in Paris; you typically spend $300 per month but the amount jumps to $10,000 in one week; you typically shop at discount stores but your card is used to make purchases at a designer clothing store.

When you respond to a report about suspected fraudulent activity on your credit card, you give information to the database. If you confirm that a digital transaction was fraudulent, the algorithm is more likely to label future transactions from the same IP address or network, or conforming to a similar pattern, as fraud in the future.

The computer algorithms can spot patterns that might not be detected by humans, and use this information to continually improve the predictions as new information accrues. These types of algorithms can detect many types of fraud (not just credit card fraud) and thus contribute to the fraud statistics produced by financial service companies and federal agencies.

CRIME INFORMATION FROM BIG DATA?

Massive amounts of data are produced as a byproduct of online activities. These include social media data, internet search patterns, comments on websites, cell phone records, and citizen requests for government services.

Can such data be used to improve crime rate estimates, or perhaps to capture crimes missed by other data sources? Research on using big data for estimating crime is still in its infancy; most of the studies conducted to date have attempted to find correlations between tweeting activities, or complaints about streetlight outages, and crimes recorded by police.

One challenge in using big data to estimate crime is that big data are often conveniently available data. Statistics computed from convenient data describe that particular data set but often do not apply to the population of interest. Data from social media and internet search patterns are volunteered, and posts and searches can be dominated by a few active persons or organizations. Large parts of the US population do not participate in social media.

Volunteered data sets also have another source of error: some of the data might not be real. A 2018 Pew Research study estimated that two-thirds of tweeted links to popular news websites were posted by automated accounts, not by human beings. More than one-fifth of those links came from a small set of 500 suspected "bot" accounts.

Margins of error are rarely reported for statistics calculated from big data. That does not mean they are error-free. As we discussed in Chapters 4 and 5, biases in these statistics from missing data and measurement error can be far greater than the sampling error from a smaller yet carefully conducted probability sample. Any statistic, regardless of source, should be evaluated based on its statistical properties—including bias.

Although it seems unlikely that big data can replace surveys such as the NCVS for obtaining accurate statistics about crime (at least not in the foreseeable future), nontraditional data sources can sometimes provide information not available elsewhere, as seen in the following example.

Crimes in Hotels

Xi Leung and colleagues extracted comments mentioning crimes from a website that solicits hotel reviews. They found that theft, fraud, and burglary were more commonly mentioned by guests at 4- and 5-star hotels, while prostitution and drug crimes were more commonly mentioned by guests at 1- and 2-star hotels. The reviews mentioning crime received more "this review was helpful" votes from subsequent website visitors, on average, than matched reviews not mentioning crime.

The study authors acknowledge that their sample is not representative of persons experiencing or witnessing hotel crime. It cannot be used to estimate the number of crimes occurring at hotels (that statistic can be obtained from the NCVS). But their exploratory study of online reviews provides information about possible relationships between type of hotel and type of crime not available from the UCR or the NCVS, and suggests avenues for future confirmatory research.

At the time the reviews were extracted, review comments about crime were not used in the website's hotel rankings (except insofar as crime victims, a tiny percentage of all reviewers, tended to give

lower scores). If online travel agents started ranking hotel security by the number of comments mentioning crime, providing an
incentive to obtain favorable numbers, the voluntarily contributed
comments might lose their value for exploring patterns. It would
take only a few fictional reviews to push a hotel into a high-crime
category.

DATA-DRIVEN POLICING

Homicide in Cali, Colombia

In 1993 the city of Santiago de Cali in Colombia had one of the
highest homicide rates in the world—104 homicides per 100,000
population, more than ten times the US rate. Mayor Rodrigo Guer-
rero Velasco, an epidemiologist by training, asked: Could data and
statistical analyses provide information for reducing homicide in
Cali?

A team of police officers, public health officials, and statisticians discovered that 80% of Cali's homicides were committed
with firearms. Two-thirds occurred on weekends, especially on payday weekends. These discoveries spurred the team to look at the
blood alcohol levels of the victims (most offenders were not caught),
which revealed that more than half had been intoxicated.

The conventional wisdom had been that most of the murders
were directly related to cocaine trafficking, but the data said that
a typical murder victim was a young man who had been drinking
late at night after payday.

The city restricted alcohol sales and suspended permits for car-
rying weapons on days that were deemed to be high risk for homicide, added police to the streets, upgraded police equipment and
education, and opened youth centers. By 1997 the homicide rate
had dropped to 86 per 100,000 population.

Was the reduction in homicides due to the reforms? After all,
homicide rates dropped throughout Colombia during the same time
period. Maybe the homicide rate in Cali would have gone down on
its own without the new programs.

To conclude that an intervention was successful, you need to
be able to estimate what would have happened if you had not implemented it. But you cannot go back and see what would have
happened under an alternative timeline. The best you can do is
to compare intervention areas with similar areas that did not get
the intervention. Statisticians recommend conducting a random-

ized experiment—randomly assigning some study areas to have the intervention and others to use the standard practice—to ensure that the intervention and nonintervention areas are similar.

The Cali program did not use a randomized experiment. But, because of budgetary constraints, the firearm restrictions were implemented intermittently during 1993 and 1994. This meant that there were time periods with no intervention that were likely similar to the periods with the intervention—similar on everything except the firearm restrictions. Researchers found that the risk of homicide was 14% lower during the periods with the firearm restrictions than during the comparable no-restriction periods. Additional evidence of the program's effectiveness came when a later mayor dismantled it, and homicide rates promptly climbed.

The program in Cali is an example of data-driven policing: using data to suggest actions that could reduce or prevent crime, and then relying again on data to evaluate effects of those actions.

A Cycle for Data-Driven Policing

New York City pioneered data-driven policing methods in the US in 1994 and today most large police departments rely heavily on data for allocating police resources and measuring progress toward goals. Figure 10.1 shows how a data-driven policing cycle uses the ideas of continual quality improvement described in Chapter 3.

Collect Data

The police department database of crime complaints and arrests usually serves as the primary source of data.

But "big data" methods often involve combining data sources, and other sources can include:

- Locations and information about retail stores, schools, parks, health clinics, restaurants, bus stops, ...

- City and state agency data about foreclosures, building code violations, vacant or abandoned buildings, causes of deaths, public health, transit ridership, 911 call records, ...

- US Census Bureau statistics about neighborhood characteristics: demographics, percent under the poverty line, percent unemployed, percent with college degrees, ...

- Data from social media, sensors, automated license plate readers, drones, surveillance cameras, ...

FIGURE 10.1 A cycle for data-driven policing

- Data purchased from data brokers who compile information about individuals' finances, health, purchases, pet ownership, reading habits, online searches, ...

Analyze Data

The data analysis does not have to be complicated. Many of the insights from the Cali data came from looking at crime patterns on a map and calculating percentages.

Typically, though, because of the large amounts of data involved, police departments use computer algorithms to predict places and times expected to have higher risks of crime. Some also predict the risk that individual persons in the city will perpetrate or be victimized by violent crime.

Some of the statistical algorithms used for making predictions have been around for more than a century, such as the regression models taught in introductory statistics classes. Others rely more heavily on computer power and have technical-sounding names such as "support vector machines" or "Bayesian networks."

Regardless of the algorithms' details, the idea behind them is simple: Find patterns and factors associated with crime incidents in the data. Then, assuming the same relationships hold in the future, predict a higher risk of crime where those factors are found.

That is exactly what was done in Cali. The crime team found a pattern of homicides on payday weekends with alcohol use and predicted these times and circumstances would continue to be higher risk in the future.

A large data set has so many possible relationships that risk predictions can depend on the interactions of hundreds of factors. The computer stores these relationships, and input data for a new location are run through the computer to get the predicted risk. Predictions can be continually updated as data accrue so that the computer "learns" from the data.

Plan Actions

The next step is to act on the predictions and plan interventions that attempt to reduce or prevent crime. Cali implemented firearm and alcohol restrictions and increased the size of the police force. Many police departments allocate more resources to locations predicted to have higher risk.

Implement Actions

The final step is to implement the actions that were planned. The implementation should be designed so that its effects can be evaluated, through conducting experiments, collecting more data and repeating the cycle.

Statistical Issues in Data-Driven Policing

The collection and merging of data for data-driven policing have raised concerns about civil liberties and privacy. The references listed on page 154 discuss these important issues, which are beyond the scope of this book, in detail.

For this book, focused on measuring crime and interpreting crime statistics, we ask: How does data-driven policing affect crime statistics and interpretations? First, though, let's take a deeper look at what it means when an algorithm is said to "predict" crime.

Probabilities Are Not Prognostications

In Philip K. Dick's short story "The Minority Report," three precog mutants predict crime before it occurs, with names of the criminal and victims and the exact time, place, and circumstances. The fictional Precrime Division rounds up future felons and places them in detention camps—not for crimes they have committed but for crimes that they were predicted to commit.

Dick's story is science fiction. Computer algorithms used in data-driven policing do not foretell the future; they identify patterns and correlations in the data. If the data indicate that recent motor vehicle thefts have been more prevalent on cold weekday mornings in apartment building parking lots, the algorithm predicts a higher probability that a future motor vehicle theft will occur under those circumstances.

But probability, even high probability, is not destiny. Patterns observed in past data do not have to carry over into the future. If more apartment-dwellers stay with their vehicles while warming them up instead of leaving them unattended, the prediction of car thefts based on earlier patterns will be wrong.

The predicted probabilities may also be of poor quality. They are only as good as the data that are fed into them, and the data sources often have errors. If the algorithm predicts that friends of known gang members are more likely to be involved in shootings, a young man who is erroneously listed as a gang member acquaintance may be watched more closely without cause. If the input data have patterns reflecting discrimination, those patterns will be propagated to the predictions.

Even if the input data were perfect, an algorithm may still give poor predictions. You can find patterns in any data set if you look hard enough, just like you can see animal shapes in clouds, but that does not necessarily mean the patterns reflect real relationships. It is important to evaluate the quality of the predictions to make sure they are not just picking up random fluctuations.

Self-Fulfilling Prophecies?

Computer algorithms cannot foretell the future. But there has been concern that data-driven policing could lead to a "feedback loop" where the algorithm predictions become self-fulfilling. Table 10.1 shows how a feedback loop can work.

TABLE 10.1   A Possible Feedback Loop in Data-Driven Policing

1. High crime is reported in a neighborhood.
2. The algorithm associates that neighborhood and/or its characteristics with higher risk of crime.
3. Based on the algorithm's predictions, more policing resources are assigned to the neighborhood.
4. Police officers, by having a greater presence in the neighborhood, detect more offenses there, leading to higher crime reports and a repetition of the cycle.

Such a feedback loop can lead to some groups being more likely to be arrested for crimes than others, and the effect can be magnified if results from the data are used to incentivize or measure police productivity, as discussed in Chapter 3. If narcotics investigators are sent to the inner city but not to the gated community, then persons with drug offenses in the inner city are more likely to be stopped and arrested. These arrests reinforce the previous prediction that drug offenses are more common in the inner city.

Crime Statistics With Data-Driven Policing

How does use of data-driven policing affect how we interpret crime statistics from law enforcement agencies and the UCR?

It doesn't. That is to say, police crime statistics should be interpreted the same way when data-driven policing is used as when it is not used.

Reread the feedback loop in Table 10.1, substituting the words "police supervisor" every time you see "algorithm." The logic works exactly the same.

Policing has always relied on judgments and predictions to allocate resources. The only difference with data-driven policing methods is that computers and algorithms help inform those judgments. The end result, whether computers are used or not, is statistics that capture only the crimes that are known to the police. A decreased assault rate might occur because there were fewer assaults, or it might occur because fewer assaults were reported.

The effects of data-driven policing need to be evaluated the same way as other interventions: through experiments in which some areas use the methods and similar areas do not, and through comparison with other data sources such as crime surveys.

Crime rate estimates are easy to calculate from the data collected. So are other numerical metrics such as average time to respond to a call and arrest rate. A community's trust in police, perceptions about the quality of police service, feelings of safety, and quality of life—all important aspects of a department's mission to protect and serve—are more difficult to measure. Surveys may capture some aspects of community views, but not everything can be quantified.

SUMMARY

More data on more aspects of life are available than ever before. This chapter provided examples of how large data sets, often collected for other purposes, can be used to provide information about crime.

In some cases, these new data sources can provide valuable information. But they all need to be evaluated with respect to their statistical properties, using the criteria outlined in Chapter 1. What types of measurement error are present, and how do they affect the estimates? What parts of the population are missing from the data?

Data-driven policing makes use of the large amount of data collected by police departments to identify patterns associated with crime. The data collection can be viewed as one step in a cycle for quality improvement.

Statistics from police departments using data-driven policing are interpreted the same way as other police statistics. All police statistics capture only the crimes known to and recorded by the police department.

CHAPTER 11

Crime Statistics, 1915 and Beyond

A T THE BEGINNING of the twentieth century, there were few reliable statistics about crime. Most Americans got their impressions about crime from newspaper stories, and sensationalist stories of crime waves sold papers. The popularity of detective fiction contributed to the perception that crime was everywhere.

The Uniform Crime Reports (UCR) and the National Crime Victimization Survey (NCVS) were launched in the 1930s and 1970s to provide data about the amount of crime. These data collections reflected the priorities, technology, and statistical capabilities of the times they were launched. What types of data would be collected today if one designed the system anew?

Chapter 1 asked: Where do crime statistics come from, and how can you tell whether they are accurate? This chapter briefly reviews the history of US crime statistics and looks at some possible ways of improving them.

Let's start with the story of a groundbreaking statistical investigation of crime in Chicago, which set out principles for collecting and using data to study crime.

STATISTICS RELATING TO CRIME IN CHICAGO, 1915

In the 1900s Chicago's many newspapers, filled with stories of murders, rapes, thefts, and assaults, competed to feed its reputation for criminal activity—a reputation (possibly undeserved) that long predated Prohibition and Al Capone's 1920 arrival in the city. The

Chicago *Tribune* declared in 1906 that "a reign of terror is upon the city ... no city in time of peace ever held so high a place in the category of crime-ridden, terrorized, murder-breeding cities as is now held by Chicago."

In 1914, the Chicago City Council decided to adopt a data-driven approach to the crime problem. The first step was to attempt to find out how much crime was really occurring in the city. The Council charged its Committee on Crime to report "upon the frequency of murder, assault, burglary, robbery, theft and like crimes in Chicago; upon the official disposition of such cases; upon the causes of the prevalence of such crimes; and upon the best practical methods of preventing these crimes."

The Committee engaged Edith Abbott (Figure 11.1), the assistant director of the Chicago School of Civics and Philanthropy, as its statistician. Abbott had earned a PhD in Economics in 1905 and by 1914 had published more than 20 books and articles on her statistical investigations about women and children in the labor force. As a resident of Hull House, the Chicago social settlement founded in 1889 by Jane Addams, she collected statistics about and organized social services for the community. She later went on to found the School of Social Service Administration at the University of Chicago, where in 1924 she became the first female dean of an American graduate school.

Abbott's 1915 report, titled "Statistics Relating to Crime in Chicago," went far beyond the standard statistical summaries of the day. She investigated the quality of the data sources, commented on how to interpret the statistics, and made numerous recommendations—still relevant today—on how to obtain better statistics about crime.

Crime definitions and counting rules matter when comparing statistics. Abbott reported that Chicago's felony arrest rate was higher than rates in New York and London, while the conviction rates were similar for the three cities. But she wrote that the comparisons were "subject to reservations" because the crime definitions, laws, and procedures varied.

Multiple data sources are needed to study crime and evaluate the quality of crime statistics. Abbott gathered data from the annual reports of the Police Department, Municipal Court, the Adult Probation Office, and the House of Correction. She also ob-

FIGURE 11.1   Edith Abbott (Public domain)

tained the unpublished reports of criminal complaints from the police department and argued that reliable statistics of these should
be kept and published each year; although "much crime is undetected," it is "most important to have statistics showing the crimes
known to the police."

   She contended that statistics about arrests and convictions,
though valuable, cannot be used to compare crime rates across
localities or times, because "one city may be very lax about ap-
prehending criminals, another very thorough and still another very
active in making arrests, but very inefficient in arresting the right
persons." Similarly, "a small number of convictions may be due
to the fact that the police have been inefficient in one city or the
courts inefficient in another."

Obtaining more accurate statistics requires changing the system of data collection.   After a thorough analysis and comparison with other sources, Abbott concluded that the statistics of
crimes known to the police, called criminal complaints, were of

questionable value. They severely undercounted the amount of crime—for some types of felonies there were more than four times as many arrests as complaints. Moreover, the ratio of arrests to complaints varied widely across precincts.

She recommended that the police department establish a uniform system for recording crimes: the common practice of "desk sergeants writing verbal complaints on slips of paper, placing them on a spindle and tearing them up when an officer reports thereon" did not lead to accurate statistics.

Accurate statistics are needed to dispel myths about crime. Abbott's report addressed the then-widespread belief that the foreign-born (who in 1910 Chicago were predominantly from Germany, Eastern Europe, and Ireland) were disproportionately responsible for crime in the city.

The Chicago *Tribune* summarized the popular view: "Chicago is infested by gangs of hoodlums to whom the law is a thing to mock at and by whom the revolver and bullet and the strong arm of the officer are the only things that are feared, and who, after dark in every locality, and through the day time in many localities, menace the decent citizen's life and belongings at every step."

Abbott, however, used data to investigate the issue. She compared the percentages of native- and foreign-born men arrested and convicted of crimes in 1913, from Chicago police statistics, with the percentages of native- and foreign-born men in Chicago from the 1910 Census. She reported that while the foreign-born accounted for 36% of arrests and 35% of convictions, they represented 54% of the Chicago population of men age 21 and over. Abbott concluded that the "various foreign groups show almost uniformly a smaller percentage of convictions than their proportion of the population entitles them to have."

Statistics need not be perfect to be useful. Abbott knew her data had deficiencies. She wrote: "The question as to how far these statistics of 'nativity' are trustworthy must, of course, be considered. In general, the method of having information about country of birth hurriedly entered by a police officer at the time of an arrest or an arraignment would undoubtedly result in many errors."

She also knew that missing data—crimes that did not come to the attention of the police or in which no one was arrested—affected her conclusions and prefaced them with the qualification:

"If the number of arrests indicates the extent of crime . . . ."

But her conclusion about crimes committed by the native-born and foreign-born was based on a careful analysis of the best data she could find. The editorials reaching different conclusions were based predominantly on personal opinions and impressions from newspaper reports.

Edith Abbott's report was one of the earliest high-quality systematic evaluations of US crime statistics. She scrutinized the quality of each crime data source, and catalogued their errors and deficiencies. Her recommendations for collecting statistics on crimes known to the police set the stage for the development of the UCR system in the 1920s.

CRIME STATISTICS, 1915–2018

Edith Abbott's call for better statistics was echoed by social scientists, police and legal professionals, and statisticians throughout the 1920s, when Prohibition and other societal changes led to increased, or at least to a perception of increased, crime. Attorney Clarence Darrow commented on the sad state of national crime statistics in 1926. He wrote that "no intelligent person can examine the statistics which are at present available and come to any satisfactory or defensible conclusion as to the number of crimes committed in the United States."

Origins of the UCR.   The committee charged with developing the UCR in the 1920s faced an enormous task: to establish order and uniformity from thousands of law enforcement agencies with their various recordkeeping systems and criminal codes. They were establishing a data system where none had existed before, and, without statutory authority to collect data, they tried to give law enforcement agencies reasons to participate and to make data submission easy.

The UCR committee emphasized that a national set of statistics would benefit all law enforcement agencies by showing trends and patterns in crime. These would enable "scientific police management" and lead to more accurate public perceptions of police activities: "In the absence of data on the subject, irresponsible parties have often manufactured so-called 'crime waves' out of whole cloth, to the discredit of police departments and the confusion of the public concerning effective measures for reducing the volume of crime."

Understanding "the reluctance of some police forces to compile and publish reports showing the number of crimes committed," they argued that crime statistics do not measure a police department's effectiveness any more than the number of influenza cases measures a health department's effectiveness.

Limiting UCR statistics to crime counts for the crimes listed in Table 3.1 made it easier for agencies to report data. Many law enforcement agencies had no experience in keeping statistics, and the manual gave detailed instructions for how to tally crimes with groups of four vertical lines crossed by a fifth diagonal line.

The committee recommended that larger cities store detailed information about crimes and offenders on punch cards: "With the aid of automatic tabulating and listing devices, the facts punched on the cards can be tabulated in any manner desired. The machines are almost human in their accomplishments; they list grand totals, totals, and subtotals for any fact or combination of facts."

The procedures and set of crimes specified for the UCR Summary Reporting System were innovative and practical in 1930. Once established, however, they set the structure for subsequent crime statistics. Although the UCR and NCVS have undergone refinements and revisions to improve the accuracy of statistics over the years, the constraints set in 1930 still affect them. When the UCR rape definition was changed in 2013, this marked "the first time in the more than 80-year history of the UCR Program" that the FBI had changed the definition of a major offense.

Origins of the NCVS. In the 1960s, like the 1920s, there was a perception that crime had greatly increased, particularly in urban areas. A commission established by President Johnson concluded that, because law enforcement agencies had widely variable statistics-keeping habits and because many crimes were unreported to the police, it was "very difficult to make accurate measurements of crime trends by relying solely on official figures."

The NCVS was established, in part, to provide estimates of year-to-year changes in crimes against persons and households that avoided the shortcomings of UCR estimates. It therefore included roughly the same set of crimes as the UCR—those in Table 3.1, with the subtraction of homicide and the addition of simple assault.

The commission strongly recommended developing methods to measure "embezzlement, fraud, and other crimes against trust" but these crimes were not included in the initial NCVS. It started measuring identity theft in 2004 and fraud in 2017.

The UCR and NCVS are both scheduled to be updated in the early 2020s. The FBI plans to retire the Summary Reporting System and switch over to the National Incident-Based Reporting System (NIBRS) in 2021. Plans for the NCVS include a redesigned questionnaire and possible changes to the NCVS sample selection and interviewing methods.

IMPROVING CRIME STATISTICS

This book has discussed some of the sources of crime statistics available in 2018. As we saw in Chapter 10, the data world is changing, and different sources will likely be available in the future. But the same principles for evaluating statistics will hold no matter what types of data are used.

Crime Definition and Classification

The US has sources of relatively high-quality data for the crimes in Table 3.1.

Statistics for other crimes are harder to come by. As we saw in Chapter 9, the limited statistics available about fraud likely miss much of it. Even fewer statistics are published on offenses such as price-fixing, corporate fraud, money laundering, insider trading, bribery, public corruption, polluting air and water, spreading computer viruses, and emerging types of cybercrime.

The NCVS and UCR do not capture these crimes. At present, the primary measurements of most of them come from enforcement activities: the statistics not only underestimate the crimes, but vary with how many resources are allocated to enforcement.

Many of these crimes were unheard of when the UCR classification system was developed. The first step toward measuring them is to define them.

A new crime classification system.   In 2015, the United Nations Office on Drugs and Crime endorsed a new crime classification system. A 2016 National Academies of Sciences panel recommended its adoption, subject to minor changes, for statistical measures of crime in the US.

TABLE 11.1   International Classification of Crime for Statistical Purposes

1. Acts leading to death or intending to cause death
2. Acts leading to harm or intending to cause harm to the person
3. Injurious acts of a sexual nature
4. Acts against property involving violence or threat against a person
5. Acts against property only
6. Acts involving controlled psychoactive substances or other drugs
7. Acts involving fraud, deception or corruption
8. Acts against public order, authority and provisions of the State
9. Acts against public safety and state security
10. Acts against the natural environment
11. Other criminal acts not elsewhere classified

SOURCE: United Nations Office on Drugs and Crime, *International Classification of Crimes for Statistical Purposes (ICCS), Version 1.0*, p. 14.

Each offense is classified into exactly one of the 11 main categories listed in Table 11.1, which encompass all of the offenses measured by the UCR and NCVS as well as new and emerging types of crime.

Every main category has multiple subcategories. For example "Acts against property only" includes burglary, theft, intellectual property offences, and vandalism, and each of these subcategories has further divisions.

Additional variables capture information on other characteristics of each offense, including information about the event (completed or attempted, type of weapon used, type of location, whether it was a cybercrime), victim (demographic information, employment, profession), or perpetrator (demographic information, relationship to victim, drug or alcohol use, repeat offender).

The additional variables would allow the type of tabulation "in any manner desired" that the 1929 UCR committee had dreamed about. Some of these tabulations, such as by type of weapon, can be done with NIBRS and NCVS data, but a uniform recording system would make calculating and comparing statistics easier.

Improving accuracy of crime classification.   Chapter 3 discussed crime misclassification in law enforcement statistics, and described some of the steps that Los Angeles took to reduce misclassifications of assaults. The Los Angeles study concluded that most of the classification errors were systems-level problems.

Because the errors are caused by systems, systems-level solutions are needed to reduce them. A system that makes it easier to place crimes in the correct category will improve classification accuracy.

Aggravated assault has been especially prone to classification errors. The UCR definition of aggravated assault includes everything from waving a gun around in a bar to nearly killing a person. It is understandable that law enforcement agencies would find this broad range confusing or might not want brandishings counted in its statistics on serious crimes.

The National Academies of Sciences Panel on Modernizing Crime Statistics recommended removing the category of aggravated assault from crime classifications. Instead, they suggested separating assaults, which can result in injury or harm, from threats. An assault would then be categorized as severe or minor based on the level of injury.

Table 11.1 uses behavioral descriptions rather than legal specifications to classify offenses. Behavioral descriptions apply across all types of crime data collections. It "avoids issues created by legal complexities, resulting in a simplified and globally applicable classification" with fewer ambiguities.

A modernized crime classification system such as that in Table 11.1 may facilitate placing crimes in the correct category and thus improve accuracy. It is easier to classify offenses by behavior and actions than by legal definitions and intent.

Using Multiple Data Sources

All statistics have errors. If data sources with different collection methods and different types of errors all produce similar statistics, however, that increases one's confidence in the findings. We saw this in Chapter 2, where the CDC and UCR statistics on homicide provide confirmation of trends in homicide over time. They also allow investigation of divergences between the two sources.

The NCVS and UCR serve as independent sources of information about crime. Figure 4.3 showed a general concordance between the police-reported serious violent crimes in the NCVS and the UCR over time, despite different crime definitions and populations included. As the NCVS produces more estimates for individual states, and NIBRS coverage expands (allowing one to extract the types of crimes measured in the NCVS), it will be possible to examine the consistency of estimated police-reported crime across the two sources in more detail.

The CDC, UCR, and NCVS statistics represent complementary, not competing, information on crime. NIBRS provides information on offenders and the relationship between offender and victim. The NCVS data provide the victim's perspective, and ask about the financial, physical, and psychological consequences of crime to the victim. Other surveys have asked people about offenses they have committed, to study why people commit crimes.

One possible model for future data collection is to supplement a large survey such as the NCVS with smaller surveys that collect data on specific topics or for smaller regions. A city, for example, might want to take a mail survey similar to the fictional example described in Chapters 5 and 6, to get a better idea of local experiences and attitudes.

New data sources, as suggested in Chapter 10, may be developed to give insight into aspects of crime not studied in the NCVS and UCR. Each data source captures a different piece in the mosaic of information about crime. Combining and contrasting statistics from different sources gives a fuller picture than using one source by itself.

Sometimes one can link individual records from the data sources. The next section describes how combining the information from the various records collected about a homicide provides information not available from a single data source.

National Violent Death Reporting System

The death certificate has information on a homicide victim's age, sex, profession and employment status, cause and time of death, and pregnancy status. The medical examiner's report describes wound locations, health conditions, weapons used, and circumstances relevant to the death. A toxicology report tells whether drugs or alcohol were present in the victim's system. Law enforcement reports contain narratives about the circumstances of the death, interviews with witnesses, and information about perpetrators or suspects.

The National Violent Death Reporting System, administered by the CDC, links the information from these data sources for each death due to homicide or suicide. By relying on information from all of the sources, persons using the combined data may be able to classify homicides more accurately (for example, as murder or justifiable homicide), and thus obtain more accurate annual homicide counts.

Linking the data records from several sources allows researchers to study characteristics of homicides that could not be investigated from a single data source. For example, an analysis of 10,000 homicides of adult women revealed that about 25% of victims did not have a high school diploma or equivalent (compared with about 13% of adult women in the population). Of the women murdered by an intimate partner, about 11% had experienced some sort of violence in the past month; about 30% of the murders had been preceded by an argument.

Even though the data are not yet nationally representative (27 states participated in 2018), these statistics suggest actions that communities, medical professionals, and law enforcement agencies might take to reduce violent deaths. Effects of the proposed actions can then be tested through randomized experiments as part of a continual quality improvement program.

Continual Quality Improvement of Crime Statistics Systems

Crime statistics are products of the systems used to collect and process the data. Improving the quality of those statistics is a continual cycle of identifying opportunities for improvement, testing proposed changes in studies or experiments, reviewing the results, and implementing actions based on what was learned.

Data collection.   In light of the discussion in Chapter 3 on incentives, data collection systems themselves can have unintended consequences for the quality of statistics. A system that is too burdensome can discourage participation or lead to worse data quality.

Deming wrote: "Data are not taken for museum purposes; they are taken as a basis for doing something. If nothing is to be done with the data, then there is no use in collecting any. The ultimate purpose of taking data is to provide a basis for action or a recommendation for action."

Data collection methods and systems benefit from ongoing quality improvement efforts in which new data collection methods are tested and evaluated.

Assessing and improving accuracy. The distinguishing feature of the discipline of statistics is the fundamental idea that every statistic should be accompanied by a measure of its accuracy. The margin of error measures just one type of error, the sampling variability that results from taking a sample instead of questioning everyone in the population.

In large surveys, and in the UCR and other "big data" sources, errors from measurement methods and missing data are far larger than errors from sampling variability. Failure to mention the uncertainty from those errors does not mean it does not exist.

Experiments and other statistical analyses can point to ways of assessing and reducing measurement error and effects of missing data in would-be censuses as well as in surveys.

Transparency. Part of producing accurate statistics is providing convincing evidence of their accuracy. Every statistic should be backed by a full methodology report, with details of how the data were collected and the statistic calculated. When possible, data should be made available to the public, subject to protecting the privacy and safety of persons in the data set.

SUMMARY

Edith Abbott, reviewing the crime statistics available in Chicago in 1915, outlined how they could be improved. Her recommendations—to have consistent crime definitions, use multiple sources of data, and improve systems for collecting data—apply equally well in 2019.

Statistical quality improvement methods can be applied to crime statistics systems to increase statistics' accuracy and utility. As Abbott said, "the importance to human welfare of a careful examination by the community of its statistics of crime is scarcely less than the study of statistics of mortality and morbidity."

We judge statistics by the procedures used to collect the data and calculate the estimates. Studies that use poor procedures rarely evaluate and report sources of errors, but they still have them—and such procedures often yield poor estimates. Good statistics, accompanied by measures of accuracy, come from sound statistical procedures.

Glossary and Acronyms

GLOSSARY

Audit: A verification system used to inspect or examine a sample of units to assess accuracy.

Bias: The tendency of a statistic to overestimate or underestimate the population value.

Census: A data collection procedure in which an attempt is made to enumerate or obtain information on all members of the population of interest.

Demographic Information: Information about characteristics of a population such as age, race, ethnicity, and sex. Sometimes includes urban/suburban/rural classification, national origin, marital status or household structure, home ownership, education or income levels, and poverty status.

Hierarchy Rule: Rule used to classify crimes in the Summary Reporting System of the Uniform Crime Reports, where each crime incident with multiple offenses is scored as the highest offense in the hierarchy. Justifiable homicide and motor vehicle theft are exceptions to the rule.

Imputation: A statistical procedure in which an estimated value is substituted for a value that is missing in a data set.

Incident: The UCR program defines an incident as "one or more offenses committed by the same offender, or group of offenders acting in concert, at the same time and place."

Margin of Error: A measure of the magnitude of error that is due to taking a sample instead of measuring the entire population.

Measurement Error: Any error that causes a value recorded in a data set to differ from the "true value."

Methodology Report: A report that describes how the study was designed, how the data were collected, and how statistics were calculated from the data.

Misclassification: Placing an item in an incorrect category of a classification system.

Nonresponse: The failure to obtain data from some units that are selected for a study.

Nonresponse Bias: A systematic difference between a statistic and the population value it estimates, caused by nonresponse.

Oversampling: Sampling a higher fraction of one population subgroup than another. If 1% of men and 2% of women are selected for the sample, women are oversampled relative to men. Sampling weights compensate for oversampling so that statistics represent every subgroup in proportion to its representation in the population.

Population: The set of persons, households, or entities of interest.

Probability Sample: A sample chosen using random selection methods, where each subset of the population has a known probability of being drawn as the sample.

Quality: The characteristics of a service or product that are related to its ability to meet societal needs. Sometimes called "fitness for use."

Representative Sample: A sample that can be used to estimate quantities of interest in a population and provide measures of uncertainty about the estimates.

Respondent: A person or household who participates in a survey.

Response Rate: The percentage of eligible persons or households who provide usable data for a survey.

Sampling Variability: The variation from one sample to another that occurs because a sample is taken instead of a census.

Sampling Weight: The number of population units represented by a unit in the sample.

Simple Random Sample: A sample that results from a random selection procedure in which every set of n units in the population (where n is a fixed number denoting the sample size) has the same chance of being selected as the sample.

Statistical Quality Improvement: A system of management principles and statistical methods that can be used to improve an organization's processes. Quality improvement never ends, as more improvements can always be made.

Stratified Random Sample: A probability sample in which a population is partitioned into groups called strata and a simple random sample is selected from each stratum.

Survey: A data collection process in which data are obtained or requested from a subset of the units in the population (as opposed to a census, in which an attempt is made to obtain data from the entire population).

Weighting Class Adjustment: A method of adjusting weights of respondents so they also represent the population share of nonrespondents in the same weighting class.

LIST OF ACRONYMS

CDC	Centers for Disease Control and Prevention
FBI	Federal Bureau of Investigation
FTC	Federal Trade Commission
MOE	Margin of Error
NCVS	National Crime Victimization Survey
NIBRS	National Incident-Based Reporting System
NISVS	National Intimate Partner and Sexual Violence Survey
SHR	Supplementary Homicide Reports
UCR	Uniform Crime Reports
US	United States

For Further Reading

CRIME HAS EXISTED as long as there have been people. Here are 12 books and reports you can read to learn more about crime statistics and deepen your understanding.

You can find the endnotes for this book, with a full bibliography of sources and additional details about material in the text, at http://www.sharonlohr.com. The same website provides links to data sources.

Crime Data

James, N. and L. R. Council. 2008. *How Crime in the United States Is Measured.* Washington, DC: Congressional Research Service. An excellent summary of some the data sources discussed in this book.

Lynch, J. P. and L. A. Addington (Eds.) 2007. *Understanding Crime Statistics: Revisiting the Divergence of the NCVS and UCR.* New York: Cambridge University Press. Reviews the history and features of the NCVS and UCR.

National Academies of Sciences, Engineering, and Medicine. 2016. *Modernizing Crime Statistics—Report 1: Defining and Classifying Crime.* Washington, DC: The National Academies Press.

National Academies of Sciences, Engineering, and Medicine. 2018. *Modernizing Crime Statistics—Report 2: New Systems for Measuring Crime.* Washington, DC: The National Academies Press. These two volumes discuss updating measurements of crime.

National Research Council. 2014. *Estimating the Incidence of Rape and Sexual Assault.* Washington, DC: The National Academies Press.

Regoeczi, W. C. and D. Banks. 2014. The nation's two measures of homicide. Technical Report NCJ 247060. Washington, DC: US Bureau of Justice Statistics.

Quality Improvement and Data-Driven Policing

Couper, D. C. 2017. *Arrested Development: A Veteran Police Chief Sounds Off About Protest, Racism, Corruption and the Seven Steps Necessary to Improve Our Nation's Police, 2nd ed.* Blue Mounds, WI: New Journey Press. Couper, who was chief of police in Madison Wisconsin for 21 years, writes about how to apply Deming's ideas of quality improvement to law enforcement agencies.

Deming, W. E. 1986. *Out of the Crisis.* Cambridge, MA: MIT Center for Advanced Engineering Study. Here is Deming's philosophy for quality improvement in his own words.

Ferguson, A. G. 2017. *The Rise of Big Data Policing.* New York: New York University Press.

Perry, W. L., B. McInnis, C. C. Price, S. Smith, and J. S. Hollywood. 2013. *Predictive Policing: The Role of Crime Forecasting in Law Enforcement Operations.* Santa Monica, CA: RAND Corporation.

Statistical Methods for Measuring Crime

These books tell how statistical methods used in measuring crime work. They assume that you have some prior knowledge of statistics. Both have numerous examples and case studies.

James, G., D. Witten, T. Hastie, and R. Tibshirani. 2013. *An Introduction to Statistical Learning.* New York: Springer. If you thought your first statistics class was just one boring formula after another, give this book a try. It introduces you to modern methods of statistics, including the methods discussed in Chapter 10 that are used in data-driven policing.

Lohr, S. L. 2010. *Sampling: Design and Analysis, 2nd ed.* Boca Raton, FL: CRC Press. Tells how sampling is done and gives the formulas, along with examples of how large federal surveys work.

Index

156 ■ Index

Printed and bound by CPI Group (UK) Ltd, Croydon, CR0 4YY

22/10/2024

01777613-0003